PRAISE FOR GEODEMOGRAPHICS FOR MARKETERS

'*Geodemographics for Marketers* is a comprehensive introduction to the what, why, when and how of geodemographics – an essential discriminator for determining a whole range of decisions, from store locations to council planning. It is a must for serious marketers and planners and will become the standard work on this subject.'
Roger Holland, Executive Chairman, JICPOPS (the Joint Industry Committee for Population Standards)

'*Geodemographics for Marketers* is that rare book on the office shelf – engaging, comprehensive, relevant and eminently practical. If you're interested in the why, how and "so what" of geodemographics, look no further. The answers are here.'
Emma White, Doctor of Mathematics, ADRC-E, University of Southampton

'Barry Leventhal has provided a timely and very readable survey of geodemographics as a marketing and commercial analysis tool, and how usage has developed over 40 years and continues to evolve in the era of big data. The value of geodemographics is clearly presented, including through contributions from a dozen leading experts steeped in relevant applications in business or public policy. This is more than an excellent "how to" guide, with its practical and usable information on techniques and data sources, because it also adds to the case for censuses of the population as a vital component of national data.'
Paul Allin, Visiting Professor in Statistics, Imperial College London

'This is a very timely book. The use of geodemographics is growing rapidly, particularly in its application in the analysis of big data sets. Under Barry's leadership, the book brings together the leading geodemographic minds in the UK to provide a clear and concise guide of how to get the best out of geodemographics.'
Lynne Robinson, Research Director, IPA

'Barry and his fellow contributors have provided a clear and very readable account of the vital role of geodemographics in marketing, demonstrating what they can achieve for the marketer, and the many ways in which geodemographic classifications can be applied. Although geodemographics have been around for well over 30 years, Barry shows that they are still extremely relevant today. Geodemographics were first applied to Kantar Media's TGI (Target Group Index) database, which I work on as a statistician today, and they remain a key tool for our clients. Marketers who read this book will be left in no doubt as to their vital contribution to successful marketing campaigns.'
Martin van Staveren, TGI Technical Director, Kantar Media

'Like Harry Beck's iconic tube map, this essential guide transforms a complicated mix of geography and statistics into an unmistakeable, coherent and actionable reference for all marketing professionals. Written to educate and inspire, this is an easy-to-follow journey through theory and practice, supported by relevant and successful case studies. Readers will put this book down with a profound understanding and the confidence that they can successfully adopt geodemographics within their business.'
Gordon Farquharson, Director of Analytics, more2 Ltd

'Geodemographics remains an essential tool for many businesses, and others are still discovering it; both need to get impartial advice. Dr Leventhal is a leading expert in this field, and it is great to see this definitive and up-to-date volume being published.'
Keith Dugmore, Director, Demographic Decisions Ltd

'An exciting journey in geodemographics, *Geodemographics for Marketers* takes the expert, and inexpert, reader on a ride from the origin of this complex discipline to its latest achievements; from the dawn of the third industrial revolution, to the spreading of the Big Data era; from the underlying theory, to relevant and interesting market applications.

'This book, including contributions from eminent exponents of the discipline, represents an unmatchable source of inspiration for everyone interested in this limitless, growing field. It is a necessary handbook of the methodologies, approaches, classifications, tools, research design, applications, potentialities and limits of a powerful scientific instrument.

'With this book, Barry Leventhal has achieved an empirically rich and academically challenging series of contributions, flavouring it with his unique experience, knowledge and market understanding.'

Claudio Calvino, PhD, Senior Applications Consultant, Open Source Data Scientist, Capgemini UK plc, and Research Associate, Oxford Internet Institute, University of Oxford

'In this era of excitement over "big" data it is important for marketers to be able to understand the data sources available to them. Without working knowledge they can neither challenge the experts nor write an accessible, decent brief. This excellent and eminently understandable book positions geodemographics as a partner to other big data streams and market segmentation techniques. It clearly and helpfully provides marketers with the information they need on this key science, and should be required reading.'

Jane Frost CBE, Chief Executive Officer, Market Research Society

'*Geodemographics for Marketers*, authored by Barry Leventhal and a star list of geodemographic analysts, is a *must*-read for students and practitioners in the field. The book provides guidance on standard and new methods, insights into "Big Data" and information on a range of commercial applications, unlocked for a wider audience by Barry's networking skills. This book is a masterpiece of co-production between commercial practitioners and academics. We can all be grateful that Mrs Leventhal asked her husband in spring 2014, "Why don't you write a book?"'

Philip Rees CBE FBA AcSS FRGS, Professor Emeritus, School of Geography, University of Leeds

'Barry has succeeded brilliantly in making geodemographics approachable and easy to understand; without leaving out anything of significance. He has covered the subject very thoroughly, and has involved some key "players" in that coverage. I believe this will instantly become the standard text for marketers interested in geodemographics.

'Barry's enthusiasm for, and deep knowledge of, geodemographics shines through. He knows his subject inside out, as well he might. He joined PinPoint Analysis in the eighties, having used geodemographics in his previous job at AGB. Since then he has been at the forefront of developments over the decades, having also started the MRS Census and Geodemographics Group in the early nineties, giving him (and the members of the group) an inside track on developments.'
Peter Sleight (practitioner since 1982 – now retired)

'This book is a comprehensive account of the development, use and future of geodemographics. In addition, it covers many related areas and initiatives, providing an extremely helpful context. The references and sources are well documented, giving access to the details that can be so difficult to find but are often critical in actually using geodemographics and fully understanding the results of having done so. Barry has to be congratulated on putting together this excellent book and having persuaded so many other leading experts to contribute. It is written with great clarity, and answers the questions that would be raised in a reader's mind just before they occur.'
Martin Callingham, former Group Market Research Director, Whitbread PLC

BARRY LEVENTHAL

GEODEMOGRAPHICS FOR MARKETERS

Using location analysis for research and marketing

MRS Evidence Matters™

KoganPage

LONDON PHILADELPHIA NEW DELHI

MARKETING SCIENCE SERIES

First published in Great Britain and the United States in 2016 by Kogan Page Limited

2nd Floor, 45 Gee Street	1518 Walnut Street, Suite 1100	4737/23 Ansari Road
London EC1V 3RS	Philadelphia PA 19102	Daryaganj
United Kingdom	USA	New Delhi 110002
www.koganpage.com		India

© BarryAnalytics Limited, 2016

The right of BarryAnalytics Limited to be identified as the author of this work has been asserted by them in accordance with the Copyright, Designs and Patents Act 1988.

ISBN   978 0 7494 7382 2
E-ISBN 978 0 7494 7383 9

British Library Cataloguing-in-Publication Data

A CIP record for this book is available from the British Library.

Library of Congress Cataloging-in-Publication Data

Names: Leventhal, Barry, author.
Title: Geodemographics for marketers : using location analysis for research
  and marketing / Barry Leventhal.
Description: London ; Philadelphia : Kogan Page, [2016] | Includes
  bibliographical references and index.
Identifiers: LCCN 2015040199| ISBN 9780749473822 | ISBN 9780749473839 (ebk)
Subjects: LCSH: Marketing research. | Geodemographics. | Consumers–Research.
Classification: LCC HF5415.2 .L48 2016 | DDC 658.4/3–dc23 LC record available at
http://lccn.loc.gov/2015040199

Typeset by Graphicraft Limited, Hong Kong
Print production managed by Jellyfish
Printed and bound by CPI Group (UK) Ltd, Croydon, CR0 4YY

CONTENTS

Supporting resources to accompany this book are available at the following url.
(Please scroll to the bottom of the page and complete the form to access the resources.)
www.koganpage.com/product/geodemographics-for-marketers-9780749473822

The same supporting resources are also available at the following url.
(Please register or log in to access the resources.)
www.barryanalytics.com/geodems4marketers

CONTRIBUTORS' BIOGRAPHIES

Venkat Anumula

Venkat S Anumula is datascience manager at Manning Gottlieb OMD (part of the Omnicom Media Group), a media agency in London. His main focus is on providing actionable insights using business data analytics to help answer clients' questions. He works on three key areas: customer analytics (including GIS and geodemographic profiling), advanced digital analytics and media response analytics. Most of his analysis is based on customer transaction-level data (offline and cookie level), primary research data and occasionally third-party data.

Venkat has over 12 years' experience working in the field of data analytics. In addition to agency experience, in the past he has worked as a research consultant and as an econometrician modelling supermarket data. His expertise spans across multiple sectors including retail, media (online and offline), fast-moving consumer goods (FMCG), travel and leisure, automobile, financial and government.

Venkat holds a degree in chemical engineering from Osmania University, India and an MBA in marketing and e-business from Indian Institute of Science. He guest lectures at Brunel University on applying statistical techniques in a marketing context, and co-authored two research papers in branding and marketing that were presented at Esade Business School and Brunel University.

Dr Robert Barr OBE

Robert (Bob) Barr is an academic geographer. He worked at the University of Manchester for 30 years and is a geographical information systems specialist. He spent a year at the University of California, Santa Barbara as a Harkness Fellow. While at Manchester he was a grant holder for over £2 million of grant funding from the research councils as well as government and commercial contracts. He was a co-founder in 1986 of the Manchester Regional Research Laboratory, which he spun out as a company, Manchester Geomatics (MGL), in 1999. He remains chairman of MGL.

Bob was a founder member of the Association for Geographic Information, spent 13 years on the association's council and was chairman in 2001.

Bob is a visiting professor in geographic information science at the University of Liverpool, and an honorary research fellow in the School of Environment at the University of Manchester. He is an expert member of the Advisory Panel for Public Sector Information and of the Open Data User Group. He is also a borough councillor and non-executive director of a major regional housing association. Bob was awarded an OBE for 'services to geography' in the 2008 New Year Honours.

Dr Luke Burns

Luke Burns is a lecturer in the Centre for Spatial Analysis and Policy in the School of Geography at the University of Leeds in the United Kingdom. Having worked in both industry and academia, Luke has developed firm expertise in several areas of quantitative human geography. These interests typically follow a spatial dimension and include the advanced application of geographical information systems (GIS) to socio-economic problems such as crime, health, planning, retail, transportation and environment. He also has extensive experience of geodemographics, census analysis and data modelling. Luke regularly delivers presentations at conferences/events on his work and teaches on a broad selection of analytical courses comprising undergraduate, taught postgraduate, online/open distance learning and continuing professional development. Luke also holds visiting lectureship status at Sheffield Hallam University and the University of Strathclyde.

Professor Tim Butler

Tim Butler is emeritus professor of Human Geography at King's College London where he has taught since 2003. He was previously Professor of Urban Sociology at the University of East London where he spent the previous 30 years of his career. Tim has undertaken research on London focusing on a number of issues around the way the city has changed since the 1970s. These include gentrification (and its euphemistically known relation of urban regeneration), East London, the changing nature of its middle classes and the sociology and geography of its education system and, most recently, its super rich. He also conducted a comparative study of the middle classes and social mixing in London and Paris with colleagues from Bristol and Goldsmiths in the UK and the University of Nanterre in Paris. He has also taught at Sciences Po Paris where he has been a visiting professor, including a spell as the Vincent Wright Professor in the Centre for European Studies. His research has mainly involved questionnaires and in-depth interviews but, as he describes in his contribution to this book, he has also used geodemographics and large-scale secondary datasets. He is the author of several books, the editor of a number of collections and is a member of the editorial board of the *International Journal of Urban and Regional Studies*.

Dr Peter Furness

Peter Furness is a mathematician who runs a small consultancy providing services in decision analytics, modelling and data mining. He also conducts research into data-mining methods and tools, seeking out ways to increase the benefits that clients can get from data-driven approaches.

Prior to setting up his own business in 1997 Peter worked for American Management Systems, CACI, Price Waterhouse and the National Coal Board. Whilst at CACI he helped pioneer the application of geodemographic and spatial modelling in areas such as branch location planning and customer behaviour analysis.

Peter is a member of the Census and Geodemographics Group (CGG) of the Market Research Society (MRS) and, with Andrew Hooks, was responsible for setting up the Geodemographics Knowledge Base (**www.geodemographics.org.uk**) for the CGG in 2000.

Tony Lovick

Tony graduated in Mathematics from Oxford University in 1987 and qualified as a Fellow of the Institute of Actuaries in 1994. He spent 21 years with Aviva Group before joining EMB, now Towers Watson, as a senior consultant. In 2015 he formed a new company, Love Actuarially, to become an independent contractor.

Tony is interested in innovative actuarial research and its delivery through pragmatic systems development. He uses big data analytics and data warehouse design to build applications in telematics and more general pricing and insurance applications.

Tony has undertaken a number of roles within Aviva, most recently as price optimization actuary, 'pay as you drive' actuary and head of statistics and development, in the personal lines pricing division of Norwich Union, where he led the implementation of full postcode risk cost models for motor and home insurance, pioneering the introduction of external data to Aviva rating systems.

Professor David Martin

David Martin is professor of Geography at the University of Southampton. In 1991 he published *Geographic Information Systems and their Socio-economic Applications*, which was one of the earliest textbooks to apply geographical information systems to social and economic themes. He has continued to research and publish in this field, collaborating closely with the Office for National Statistics and directing the Economic and Social Research Council's (ESRC) Census Programme from 2002 to 2012. He led development of a new automated system for the creation of output areas for the census, an approach that was successfully adopted in 2001 and subsequently developed to produce systems of super output areas, 2011 census output areas and workplace zones for England and Wales. He has ongoing interests in the construction of gridded and time-specific population models and in all aspects of geographical population referencing, most recently as a co-director of ESRC's Administrative Data Research Centre for England. David has in the past served as chair of the Royal Geographical Society's Quantitative Methods Research Group and has been an ESRC council member since 2010.

David Miller

David Miller first joined Claritas in 1981. In his current position within Nielsen's data sciences group he is responsible for innovation in global market segmentation, demographic methods, satellite imagery and market segmentation.

From 1990 on during his tenure at Claritas and Nielsen in the United States, David has been responsible for the development of all segmentation products. In particular, he led the development of the PRIZM™, ConneXions™ and P$YCLE™ segmentation systems and their evolution. During his five-year tenure in Europe he developed segmentation systems for Sweden, the Netherlands, Spain, Poland, France, Germany, Ireland and Britain.

David has worked extensively, in both the United States and Europe, in the development of intercensal demographic estimates, estimates of product sales and use as well as the development and execution of custom modelling projects. This work includes statistical models covering churn, lifetime value, customer retention and relationship management, as well as the use of satellite image information in demographic and business activity estimates.

David holds an MSc in Statistics and a Master of Philosophy in Experimental Psychology from George Washington University.

Kyle Poppie

Kyle Poppie is director in data science at Nielsen. His responsibilities include market segmentation innovation and syndicated segmentation enhancement, researching statistically sound and scalable ways to measure and target consumer segments. He has researched and implemented data integration techniques that link what consumers watch with what they buy, led methodology and assessment initiatives on numerous big data sources, and has worked extensively on creating statistical solutions that increase the utility of big data by mitigating bias and measurement error. Kyle holds an MA in Sociology from the University of Wisconsin–Milwaukee.

Mark Watson

Mark Watson is the owner of Bluewave Geographics, a leading provider of digital mapping and geographical analysis services to the market research and fieldwork sectors.

Mark graduated from Exeter University with a degree in Mathematics in 1983 and then completed a Masters degree in Operational Research at Southampton University in 1985. He went on to work in the Statistics Departments at Plessey Radar, AC Neilsen and Royal Mail.

Mark first started working in geodemographics for Pinpoint Analysis in 1992. After a brief spell at MORI, where he set up an in-house sampling department, he then spent 10 years as a director and partner at Business Geographics, an innovative digital mapping and geodemographics agency.

Mark has since continued to provide survey sampling and digital mapping consultancy and founded Bluewave Geographics in 2011. A hands-on analyst, he has many years' experience of project-based work in the areas of survey design, sampling, web-mapping and geodemographics. He has also developed a number of web-based mapping and profiling systems for Bluewave Geographics for applications in the market research industry, and has also made use of the rise in freely available open data to create a number of data products.

Professor Richard Webber

Richard Webber is the pioneer of the use of geodemographic classifications in target marketing, having originated Acorn whilst at CACI and Mosaic whilst at Experian. After graduating in Economics at Cambridge University and a Master's degree in Transport Planning at the University of Liverpool, Richard joined the Centre for Environmental Studies in 1973, developing computer algorithms that classified the different types of neighbourhood in Liverpool appropriate for different area-based policies.

The Office of Population Censuses and Surveys then commissioned Richard to create a national classification of wards and parishes. It was the adoption of this classification by the British Market Research Bureau as the sampling frame for the Target Group Index that enabled marketers for the first time to recognize consumption differences between different types of neighbourhood.

Since retiring from Experian Richard has developed Origins, a tool that enables consumers' ethno-cultural background to be inferred from their names. Webber Phillips Ltd uses this technology to help organizations measure levels of diversity using employees, citizen and customer files.

Richard is an honorary fellow of the Market Research Society, a fellow of the Institute of Direct Marketing and a visiting professor at King's College London.

Simon Whalley

Simon has a Master's degree in Geographical Information Systems (GIS). His specialist areas of interest are the design of geographical area to accurately represent spatial data and working with census or social data.

His principal role at Beacon Dodsworth is dealing with digital data for the company. This incorporates the creation, cleaning and analysis of data using databases and written software. He is the creator of P^2 People and Places, which is Beacon Dodsworth's geodemographic classification of the UK, primarily built using 2001 UK Census data and Target Group Index data and now being rebuilt using 2011 census data and British population survey lifestyle data. He also has expertise in address data and address matching, road networks and route navigation, geographical boundaries (both postcode- and administrative-based), digital raster map creation and social datasets such as the Indices of Multiple Deprivation and building supply and demand models.

Simon is the deputy chair of the Market Research Society's Census and Geodemographics Group, which is a leading independent voice championing the generation and use of geodemographic data.

FOREWORD

As Barry Leventhal describes in Chapter 1, the 1979 Market Research Society (MRS) annual conference was when the commercial potential for a cluster system based on census data and postcodes was first presented. In the seminal paper by Bermingham, Baker and McDonald (1979), from the BMRB market research agency, the authors described the findings generated by applying such a process to the Target Group Index (TGI) survey. Geodemographics was born – and for me, as a member of the then MRS Conference Committee, a real 'I was there' moment. I have attended most MRS conferences since 1978, but few of the many, many presentations I have witnessed over the years have left such a strong and lasting impression. It was the dawn of a new era in market research where a process replaced the need to ask questions, and a new, insightful, method for profiling and segmenting consumers had arrived.

However, this was not my initial exposure to the methods underpinning geodems, as I had already read a fascinating article by a then relatively unknown Richard Webber in *New Society* describing his initial research using census data to identify areas of deprivation in the Wirral area of north-west Cheshire and Merseyside. Today, as Richard, the father of geodemographics, describes in the book, it remains as useful a tool for marketers now as it was back in 1979. Geodems codes are still applied to TGI data and, as Barry describes, are added to all the main continuous surveys used to track media and product consumption. Geodems has not proved to be simply a fad; it has endured, and been continually refreshed to make it relevant in a rapidly changing world. Underlining this continuing relevance, in 2014 the BBC broadcast a series on Radio 4 describing the evolution and application of geodems in commercial and public-sector fields.

Barry, whom I have known and respected for many years, has been at the forefront of geodems since the early days. He was instrumental in setting up the MRS Census and Geodemographics Group over 25 years ago, with its extensive online resources and annual conference. He has also contributed extensively to the Institute of Direct Marketing's education programmes and their publications. Who better to argue so persuasively the value that geodems offers to marketers?

Two key factors were necessary for the successful development of geodems. First was the Royal Mail's drive to apply postcodes to all properties in the UK and develop the Postcode Address File (PAF) product. Second, the increased access to UK Census data from the mid-term census in 1966, and especially after the main one in 1971 (the first stirrings of the 'open data' revolution?). Whilst the census remains the key source of data on the UK population, it is only updated every 10 years, which in today's fast-paced, dynamic world is no longer as acceptable as previously. Geodems providers are increasingly down-weighting the importance of this source, using 'open data' sources of public data and their own commercial databases to create a base that can regularly be updated, thus creating families of products for different applications, creating added value to users, and providing agencies with enhanced revenue opportunities. But the principles that underpinned the first generation of products remain the same.

One of the great benefits of geodems as a segmentation tool is the fact that it can be easily and universally applied in targeting marketing activity. Barry mentions the alternative values-based classifications, such as Values, Attitudes and Lifestyles (VALS). I have worked with all the main ones, but whilst they provide very interesting insights from a research perspective, unlike geodems they can be very difficult to apply as a practical segmentation tool to support operational marketing. This also applies to other bespoke segmentations.

As described in the book, geodems has an increasing international reach. These systems often require some very creative thinking for their development in countries where population data is either unavailable or unreliable, and with no property identification system. Another difference that geodems clearly identified is the 'inside out' nature of cities in emerging economies where the wealthy inhabit inner-city areas and those at the bottom of the pyramid live in townships on the outskirts, the opposite of tenure patterns in many western cities.

Barry describes the relationship between geodems and 'big data', but I would argue that in an era where computing power was limited and customer databases only starting to evolve, geodems provided the first opportunity to link different data sources together in the search for a holistic profile of customers – the birth of big data in the field of marketing and the first step towards data fusion. The niche geodems-based products from this era are described in the book.

Finally, we also increasingly take 'data visualization' for granted as the best way to show patterns in complex data. I would argue that geodems was an early trendsetter in this field, as highly visual maps provided a very effective way to present the data and tell a story that users could relate to.

So, geodems still has a lot to offer marketers, as Barry and the other contributors remind those of us old enough to have been there at the beginning. This new book underlines the enduring nature of the methodologies, and how creative thinking by providers has enabled the products to remain as relevant today as they were when first developed nearly 40 years ago. I hope that the content inspires you to use geodems to explore your market and to add value to your marketing strategy.

Peter Mouncey FMRS FIDM

PREFACE AND ACKNOWLEDGEMENTS

The inspiration for this book came from my wife – 'Why don't you write a book?' asked Hazel, one quiet day in spring 2014. Normally, I ignore such flippant questions; however, when this was immediately followed by a similar invitation from Kogan Page's commissioning editor, the project suddenly became 'meant to be'.

The timing for a new book on geodemographics also felt 'right' and this gave me further encouragement. The marketplace in the UK was going through its 10-year regeneration, following the release of 2011 UK Census data, so there would be new products and systems to write about. At the same time, the National Statistician had just recommended that an online census should be carried out in 2021. So the bedrock data underpinning 'geodems' looked secure, which should fuel the industry through to the end of the 2020s.

Geodems has played an important part in my life for more than 30 years. I worked with supplier Pinpoint Analysis for five of those years, in the late 1980s. From the 1990s onwards I have mainly been engaged in analysing customer data – I never miss an opportunity to apply geodemographic discriminators and look at their effects. I have also kept in touch with the industry through the Market Research Society's Census and Geodemographics Group, which I founded in 1989.

As a benefit of writing this book, I have had to refresh my knowledge about geodemographic systems and techniques. I have found that, while many of the same core methods are still being used as I employed 30 years ago, nowadays they are applied with greater degrees of sophistication, automation and speed than I had ever imagined possible.

When I started to organize my thoughts and create a structure for the book, I was fortunate to have Richard Webber as my sounding board – I'm very grateful to Richard for his advice and suggestions.

This was followed by a research phase – I'm very grateful to the companies and people who have spent time in briefing me and providing explanatory materials about their systems.

My tour of suppliers started at CACI in November 2014, and I must thank Paul Winters for making this possible, John Rae for his briefing and Patrick Tate for producing numerous worked examples. I am particularly grateful to CACI for the series of examples used to illustrate the mechanics of geodemographics in Chapter 6.

The following month, I visited Experian in Nottingham – my thanks go to Paul Cresswell for enabling this, Matt Southgate for arranging an excellent day of knowledge sharing and for sharing materials to help with the book. I am also grateful to other Experian experts for their briefings – Ben Fenwick, Matt Holgate, Kevin Smith, Declan Mullan and Ed Cleator. Experian very kindly provided example tree outputs for Chapter 6 and also a case study for Chapter 7 – many thanks to Mark Pestereff for arranging the latter.

In January 2015 I spent a day with Callcredit in Leeds, and am very grateful to Dick Caulton for enabling this, and to Andy Peloe for arranging an excellent briefing day and for his help. Also, my thanks go to Libby Plowman, Gary Childs, Chris Duley, Chris Hill and Martin Bradbury for their briefings.

Other geodemographic suppliers have also provided briefings and materials – I am very grateful to Simon Whalley at Beacon Dodsworth and David Griffiths at TRAC Consultancy. Also, thanks go to Tim Drye and Steve Abbott for supplying a comparison example from the British Population Survey.

I am grateful to all of the industry experts who have written articles and case studies for this book. My thanks go to Richard Webber, Bob Barr, Simon Whalley, Tony Lovick, Venkat Anumula, Mark Watson, Tim Butler, Peter Furness, David Martin, Kyle Poppie, David Miller and Luke Burns for sharing their knowledge and their views. The retail case study came from Beacon Dodsworth – I must thank Karen Douglas for supplying this.

I am indebted to Luke Burns for suggesting the addition of an online practical exercise, in order to complement my theoretical description, and for creating the practical.

I must thank everyone involved at Kogan Page, especially Melody Dawes, Jasmin Naim, Jenny Volich, Philippa Fiszzon and Amanda Dackombe as well as Richard Burton. Also, I am very grateful to Keith Dugmore and Emma White for reviewing the content, and also to Kogan Page's external reviewers.

Lastly, special thanks go to my family – particularly to my brother Lionel for his wise advice about publishing, to my elder son Matthew for developing time-saving textual analysis, to my younger son Laurence for his help and to Hazel for being my inspiration.

Barry Leventhal

Publisher's acknowledgement

Kogan Page thank the Market Research Society (MRS) for their involvement with and generous endorsement of this book.

LIST OF ABBREVIATIONS

ABI   Association of British Insurers

Acorn   A Classification of Residential Neighbourhoods, a geodemographic segmentation from CACI

ADRN   Administrative Data Research Network, a UK organization that helps researchers to carry out research using administrative data

ADS   Analytic dataset (used in data mining)

BIS   Department for Business, Innovation & Skills, a UK government department

BMRB   British Market Research Bureau

BPS   British Population Survey

CDRC   Consumer Data Research Centre

CES   Centre for Environmental Studies, an environmental think tank in the UK (closed in the 1980s)

CGG   Census and Geodemographics Group, an advisory board of the Market Research Society

CRISP   Cross Industry Standard Process (for data mining)

CRM   Customer relationship management

DMP   Data-management platform

DSP   Demand-side platform

DWP   Department for Work and Pensions, a UK government department

ED   Enumeration district, the unit of geography for collecting census data in the UK

EPOS   Electronic point of sale, technology that records retail transactions

ESRC   Economic and Social Research Council, one of the Research Councils in the UK

EU   European Union

FMCG   Fast-moving consumer goods

FRS	Financial Research Survey, a survey on consumer finances in Britain
FiNPiN	Financial Pinpoint Identified Neighbourhoods, a past geodemographic segmentation from Pinpoint Analysis
GCSE	General Certificate of Secondary Education, an academic qualification used in England, Wales and Northern Ireland
GIS	Geographic information system
HMRC	HM Revenue & Customs, a UK government department
HRP	Household reference person, an individual selected as a reference point within each household (used in UK censuses)
IMD	Index of Multiple Deprivation, a measure of deprived areas used in the UK
IPA	Institute of Practitioners in Advertising, the professional body for advertising, media and marketing communications in the UK
IT	Information technology
LA	Local authority, an organization that provides local government in the UK
LAU	Local Administrative Units, the smallest geographical areas in the NUTS classification
LCF	Living Costs and Food, an ONS survey on spending patterns, food consumption and the cost of living
LOAC	London Output Area Classification, an open geodemographic segmentation for London, UK
MAUP	Modifiable areal unit problem
MRS	Market Research Society, the professional body for those working in market, social and opinion research
NAG	National Address Gazetteer, a definitive UK address list
NHS	National Health Service, a public body that delivers health services in the UK
NII	National Information Infrastructure, a collection of open public data in the UK
NISRA	Northern Ireland Statistics and Research Agency, the principal source of official statistics and social research on Northern Ireland
NMA	National mapping agency

NPD	New product development
NRS	National Records of Scotland
NS	National Statistics
NSO	National statistical organization
NUTS	Nomenclature of Territorial Units for Statistics, a classification of geographical areas for statistical outputs across the EU
OA	Output area, the smallest geographical unit for census output in the UK
OAC	Output Area Classfication, a geodemographic segmentation of output areas provided by the Office for National Statistics
ODI	Open Data Instute, a UK organization that promotes open data
ODUG	Open Data User Group, which encourages the release of open data in the UK
ONS	Office for National Statistics, the national statistical institute for the UK
OPSI	Office of Public Sector Information in the UK
OS	Ordnance Survey, the national mapping agency for Britain
PAF	Postcode Address File, a UK address list operated by Royal Mail
PCA	Principal component analysis, a multivariate analysis technique
PiN	Pinpoint Identified Neighbourhoods, a past geodemographic classification from Pinpoint Analysis
PSPP	A free statistical analysis package intended as an alternative to SPSS
QGIS	Quantum GIS, a free open-source GIS package
RFI	Request for information
RFP	Request for proposal
ROC	Receiver operating characteristic, a type of graph that illustrates the performance of a discriminator
SAR	Sample of Anonymized Records, selected from a census database
SEC	Socio-economic classification, a segmentation of people based on their occupations
Socitm	Society of Information Technology Management
SPSS	Statistical Package for the Social Sciences, a software package available from IBM

TGI	Target Group Index, a survey operated by Kantar Media
TOAC	Temporal Output Area Classification, a segmentation of output areas looking at geodemographic change
UKSA	United Kingdom Statistics Authority
UN	United Nations
UNECE	UN Economic Commission for Europe, one of the regional commissions of the United Nations
UNSD	UN Statistics Division
USCB	United States Census Bureau
VALS	Values, Attitudes and Lifestyles, a psychographic segmentation system
ZIP	Zone Improvement Plan, ZIP codes are a system of postal codes used by the United States Postal Service

INTRODUCTION

Nowadays, we learn about applications of 'big data' on a daily basis, while we rarely hear about geodemographics. However, arguably, geodemographics was the first instance of big data in marketing – in the early years of the industry, before the advent of desktop computing, large mainframes were needed for building and deploying neighbourhood classifications.

This low visibility for geodemographic applications is a reflection of the fact that many companies developed and implemented the processes some years ago, and they now have well-established functions within operational and analytical systems. Keith Dugmore, who runs the Demographics User Group, **www.demographicsusergroup.co.uk** – whose members include Argos, Boots, Camelot, Co-operative Group, E.ON, John Lewis, Marks & Spencer, Sainsbury's, Tesco and Whitbread – has stated that most, if not all, member companies have the use of geodemographics embedded in their businesses.

This book has two main purposes. The first is to brief marketing professionals on the science of geodemographics and serve as a practical guide to its many applications. Geodemographics is well established as a data-driven analysis tool for marketers. Since its launch into the UK market in 1979, the technique has been continuously used – geodemographic classifications are widely embedded in customer databases, geographic information systems and market research datasets.

Our second purpose is to explain why information about where people live is still relevant and useful, even in today's world of big data, which can provide specific factual information and individual behaviour patterns. There is really no better time for discovering, or rediscovering, geodemographics by reading this book. This claim is made for two main reasons. First and foremost, marketing processes have changed almost beyond recognition since the birth of geodemographics. For example, back in the 1970s, direct marketing communications were primarily sent by post and methods of targeting were in their infancy. Nowadays, of course, most marketing messages are delivered by e-mail, supported by online advertising. As this book explains, the science of geodemographics has evolved in parallel, supported by advances in data sources, analytical techniques and operational systems. So the data and techniques described in this book can be of value to marketing professionals who may not appreciate the role that external information can play.

Second, in the UK, geodemographic systems have recently been refreshed by a new set of results for the UK Census, as conducted by the three census offices that cover this country. At the time of writing, eight new general-purpose geodemographic classifications have been launched, along with a host of market-specific discriminators and indicators, as well as individual-level segmentations, all of which are discussed in this book.

This book follows a broadly evolutionary structure, starting with two chapters of a positioning nature. Chapter 1 summarizes the modern origins of geodemographics, overviews how neighbourhood classifications work, and places them alongside other methods of market segmentation. Chapter 2 discusses the relationship with big data, and a contribution by Richard Webber explains why the contextual information provided by geodemographics is useful to have, in combination with specific facts and behaviours about individuals.

Chapter 3 primarily discusses the ingredient data sources used in constructing geodemographic systems, including open data, which is introduced by Bob Barr, a founder member of the Open Data User Group. This chapter also describes the types of databases to which the systems are often linked.

This leads on to Chapter 4, which focuses on neighbourhood classifications – how they are built, the different types of products and the current classifications available in the UK market. The geodemographics market continues to develop at a rapid pace, with the release of new data sources. In order to future-proof this book and support readers, Chapter 4 provides details of supporting online resources containing updated information on the general-purpose classifiers.

Alongside the neighbourhood classification systems, there are many other kinds of discriminators, ranging from raw variables through to multi-variate indices, as well as segmentations that go down to household and person level. All of these other products are discussed in Chapter 5; this chapter includes a case study by Simon Whalley demonstrating the useful-ness of 'social grade', from the census, in helping to interpret Scottish referendum results.

Chapter 6 covers the 'nuts and bolts' of using geodemographics, ex-plaining the different types of profiling techniques with the help of a series of maps and profiles supplied by market analysis company CACI.

Having discussed geodemographic systems – how they are built and used – Chapter 7 focuses on describing how they are applied, starting with general principles and then providing use cases in a variety of differ-ent industry sectors. A number of external case studies and articles are used in order to bring the various applications to life. From the retail sec-tor, a case study from Beacon Dodsworth describes how one of its clients used geodemographics to prioritize alternative locations for a new store. Within the financial sector, a contribution by Tony Lovick discusses the ap-plications in insurance pricing. And in the public sector, a case study from Experian explains how Newport City Council used Mosaic to help under-stand citizens and move them over to a more cost-efficient channel. Chapter 7 closes with three contributed sections discussing the use of geodemographics for different types of research. Venkat Anumula writes about media research and includes two case-study examples. Mark Watson explains the ways in which the classifications are used for sam-pling and controlling market research surveys, and provides a worked ex-ample. And Tim Butler discusses some of the applications of Mosaic in academia, for research in social science.

By this point, any reader who is new to geodemographics will doubt-less be convinced about the benefits and will want to select the best product for their organization. Chapter 8 provides guidance on choosing a geodemographic classification, which depends on whether the supplier and system will meet the user's needs, plus how well the product discriminates in the relevant sector or market. A case study is included in order to demonstrate how discriminators may be compared, using data from the British Population Survey.

The main focus of this book is the UK market – the same principles, techniques and applications will apply to other countries, subject to necessary data sources being available. Chapter 9 discusses the international perspective, with the help of three contributed sections. Peter Furness considers the availability of censuses in different countries, while David Martin examines the small-area geographies for which census statistics are produced in other countries. Finally, Kyle Poppie and Dave Miller provide a case study on the market for geodemographics in the United States.

For those readers who are analytically minded and interested in developing their own products, Chapter 10 contains advice and guidance on building your own discriminator and creating small-area estimates for your specific market. The chapter introduces an online practical exercise, created by Luke Burns, which is designed to give readers a taste of building their own geodemographic segmentation.

Finally, Chapter 11 looks to the future and discusses plans for the 2021 UK Census, together with likely developments in open data, use of administrative datasets and the all-pervading big data.

AN OVERVIEW OF GEODEMOGRAPHICS

01

Introduction

Geodemographics is a multifaceted science. It draws on multiple sources and gives rise to a raft of products that can be employed in a variety of different applications, so there is much detail to understand. Rather than jumping straight into the mass of detail, this chapter presents an overview that is particularly designed for readers who are new to the subject area. The main aims of the chapter are:

- To explain the principles of geodemographics and illustrate its operation with a worked example.
- To identify key strengths and weaknesses of the technique.
- To identify some of the other market segmentation tools employed in marketing and discuss their relationships with geodemographics.

The chapter ends with a study that was designed to validate whether geodemographics actually works.

Definitions and principles

Geodemographics is often defined as 'the analysis of people by where they live' (Sleight, 2004). It brings together 'geo', implying geography, and 'demographics' of households or individuals into a single term.

> **KEY POINT**
>
> The level of geographical detail used for geodemographics is not defined but is taken to be small areas, such as groups of postcodes or individual postcodes. Throughout this book, the term 'neighbourhood' will be used as a shorthand term for a small area of this kind.

Two principles underpin the use of geodemographics: 1) two people living in the same neighbourhood are more likely to have similar characteristics than two people chosen at random – in other words, 'birds of a feather flock together'; 2) neighbourhoods can be categorized according to the characteristics of their residents; two neighbourhoods belonging to the same category are likely to contain similar types of people, even though they may be geographically far apart. These principles imply that geodemographic information about an area may be used to infer the likely characteristics and behaviour patterns of its residents.

Importance of census data

Even in the current era of 'big data' there are very few information sources that can provide accurate and consistent data on resident characteristics for every neighbourhood in the country. The most important and valuable such source, by far, is the population census, which is carried out 10-yearly in the UK. This collects demographics of individuals and households for a wide range of topics such as age, gender, ethnicity, religion, working status, occupation, household size, housing tenure, size of dwelling and car ownership. Therefore, the UK Census is a vital source of geodemographic information in its own right, and is also the primary input for an industry of derived geodemographic products.

Every 10 years, the release of small-area census output acts as the trigger for the geodemographic industry to 'reinvent itself' by developing a new generation of products fuelled by the latest results.

Neighbourhood classifications and other approaches

While the census is the primary source for geodemographics, the amount of detail that it provides is far too great for most users and for most purposes. The information is often 'boiled down' using multivariate analysis techniques into a neighbourhood classification, which assigns each small area to a type based upon the characteristics of its residents. By the end of 2014, eight such classifications had been launched into the UK market by various suppliers, making use of data from the 2011 census.

Neighbourhood classifications have been at the heart of geodemographics since its early days, therefore users may easily believe that the only geodemographic products are classification systems such as Acorn or Mosaic. However, there are a number of other approaches that can sometimes be more appropriate – these include analysing raw variables (eg age or household size) or creating derived indicators, such as measures of wealth. At the same time, there are also segmentations of households and individual people. Chapter 4 explores neighbourhood classifications, including the current generation of products, while Chapter 5 discusses these other approaches.

How neighbourhood classifications work

In order to illustrate how the geodemographic approach operates, we will work through a small example using a fictitious neighbourhood classification and invented data – purely to illustrate the point. For simplicity, this classification contains just six broad categories and assigns every neighbourhood in the UK to one of six segment codes, A to F. The aim of this example is to provide information on which types of people are more likely to be customers for a particular consumer product or service – or, more accurately, which categories of neighbourhoods are more likely to contain customers. The technique is known as 'customer profiling' and results in the 'customer profile' report shown in Table 1.1.

Since all neighbourhoods in the UK have been classified using our fictitious geodemographic, and the UK Census also provides the population size for every area, we are able to obtain the share of the population residing in each neighbourhood category. This information is shown in the fourth column of data in Table 1.1, ie 10 per cent of people reside in category A, 16 per cent reside in category B and so on.

Suppose that a company has 10,000 customers and plans to expand its business by finding 'lookalikes' – people who are similar to those customers – in terms of living in the same neighbourhood types. Using each customer's postcode, their neighbourhood is identified and hence their geodemographic segment, from A to F. This process is carried out for all 10,000 customers and the results are summarized, resulting in the numbers of customers shown in the second column of Table 1.1, together with the shares given in the third column. This shows that 14.5 per cent of customers belong to category A, 30 per cent to category B and so on.

The final step is to compare the two sets of shares. For example, category A has 14.5 per cent of customers coming from 10 per cent of the population, therefore this gives an index value of 145 (fifth column of Table 1.1). These index values are the key results to scan and interpret, so are often accompanied by bar graphs for helping to see the highs and lows. In this example, categories B and A are clearly the most important on index size, while customers are least likely to come from categories F, D and E respectively.

This six-category classification works well in identifying the 26 per cent of UK population belonging to types A and B that supply 45 per cent of customers. Since the classifier's main purpose is to discriminate between

TABLE 1.1 Example customer profile

Category	Number of Customers	Share of Customers	Share of UK Population	Index
A	1,450	14.5%	10.0%	145
B	3,000	30.0%	16.0%	188
C	2,000	20.0%	21.0%	95
D	1,500	15.0%	23.5%	64
E	1,500	15.0%	19.5%	77
F	550	5.5%	10.0%	55
Total	10,000	100.0%	100.0%	

Index = (% of customers / % of population) x 100

areas with high and low customer penetrations, it is sometimes known as a **geodemographic discriminator**, and categories A and B are known as **target segments**.

The company could apply this knowledge in different ways, depending on its business processes. For example, if it advertises in local newspapers, it might decide to select the towns or local authorities containing the highest concentrations of categories B and A, and target its advertising in those areas. If it sells its products through shops or stores, it might next review how well the store network matches the distribution of its customers and locate new stores on proximity to the target segments. The application of the technique varies between industry sectors, as we will see in Chapter 7. However, the concept of customer profiling underpins most of these uses. The techniques for producing and interpreting geodemographic information are described in Chapter 6, including the delivery systems available to users and the role that market research databases can play.

Key strengths and weaknesses of geodemographics

Strengths

Ability to locate

The great strength of geodemographics lies in its ability to pinpoint the geographical locations of each segment. These can be communicated either on maps or as listings of streets and addresses, depending on the user's needs. For example, one possible goal might be to identify postal sectors containing the highest concentrations of a target segment, while another might require a list of target postcodes or addresses. Systems are available that will output the target locations in a variety of alternative formats, such as maps, area listings or name and address files.

Universal coverage

The key to using geodemographics is the postcode and this has universal coverage throughout the UK. Postcodes are maintained by Royal Mail and form the basis for the distribution of post, so have complete coverage across the population. In order to apply any neighbourhood classification to customer data, a look-up file or directory is used, containing all postcodes in the UK together with their geodemographic segment codes. This method achieves high match rates and accuracy levels, provided that the customer postcodes are up to date.

Ease of use

Geodemographic systems are easy to use. For example, profile reports can be obtained simply from a file of customer postcodes. Most people tend to know their own postcode, and the postcode is often captured as the first stage in collecting a customer's address. Address verification systems will check and correct postcodes, which should therefore be obtainable to a high degree of accuracy.

Data protection

It is advantageous that the full postal addresses of customers are not required for using geodemographics at neighbourhood level, both for operational simplicity and from a data protection point of view.

In the UK, the Data Protection Act 1998 governs the processing of personal data about people. It defines personal data as any data that can be used to identify a living individual. The postcode, by itself, does not generally constitute personal data – on average, a residential (small user) postcode contains 15 addresses. However, in some circumstances (eg sparsely populated rural areas) a postcode will contain a single address and so will constitute personal data. More typically, a house number is required together with a postcode in order to define an address, but this level of detail is not needed for appending neighbourhood classifications – the postcode is always sufficient. Therefore, geodemographic analysis at neighbourhood level should not generally depend on personal data and so is relatively safe from a data protection point of view.

Contextual effects

While other information sources – notably big data – can provide specific facts about an individual, geodemographics paints a picture about the local area in which they live. This contextual information can have predictive value and is useful for descriptive purposes, alongside the hard facts. Big data and contextual effects are discussed further in Chapter 2.

Weaknesses

Not everyone is the same

Most geodemographic systems operate at neighbourhood level, classifying down to the level of individual postcodes, corresponding to streets or blocks of addresses. However, as we all know, not everyone is the same – even within a street. Therefore the geodemographic description will not necessarily fit everyone: there will always be differences and outliers. This weakness, sometimes known as the ecological fallacy, is discussed in more detail in Chapter 6.

Degradation over time

The majority of classifications rely on population census data to provide the demographic characteristics of each neighbourhood, and their accuracy will degrade during the 10-year period between successive censuses, as families mature, people move home and new households are formed. The extent of change is particularly severe for new towns and other areas where major housing developments have taken place. In general, it is much less of an issue across the majority of the UK where areas alter very gradually – even though households may move home within a decade, they will often be replaced by broadly similar households, depending on the type of accommodation and locality. Thus the change in area characteristics between censuses is less than might be expected.

Difficulty in updating

The general reliance on the population census as an input implies that geodemographic products cannot be updated for changes in area characteristics during the inter-census decade. The only way to create an updatable neighbourhood classification would be to base it solely on sources that are maintained and regularly refreshed, such as administrative databases. As we will see in Chapter 4, several of the 'post 2011' classifications have taken this approach; however, the majority continue to employ the UK Census, due to the essential importance of its demographic data.

New classifications following each population census

As we have seen, geodemographic products are highly dependent on census data; following the release of small-area outputs from each census, every supplier rebuilds its classifications using the latest available data and techniques. This creates a massive discontinuity in the market each 10 years, and implies that users may then need to recreate their in-house targeting models and systems in order to utilize the new discriminators. However, it is preferable to make use of the latest information, rather than continue to rely on a previous census that is long out of date.

Origins of modern geodemographics

Some knowledge of history often helps in understanding the present, and the same thinking is valuable for geodemographics. The following brief overview starts with earliest known use, jumps almost 100 years and then proceeds decade by decade, driven by the release of output from each decennial census. A much fuller account of the evolution of the industry is given by Peter Sleight (2004).

Earliest use

The earliest public use of geodemographics is arguably Charles Booth's mapping of London in the late 19th century, in which streets are colour-coded according to the socio-economic conditions of their residents. Examples of his maps may be found at **http://booth.lse.ac.uk/**, while his classification of streets is reviewed by Harris, Sleight and Webber (2005).

The 1970s

The origins of modern geodemographics in the UK go back to the mid-1970s and were triggered by the release of a national set of machine-readable small-area statistics from the 1971 census. Richard Webber, while at the Centre for Environmental Studies, used this data to develop national classifications of areas for research into inner-city deprivation. Subsequently, he went on to produce a national classification of wards and parishes. Ken Baker showed the value of this system for analysing the Target Group Index (TGI), a large market research survey operated, at the time, by the British Market Research Bureau (BMRB); the results were presented at the Market Research Society Conference by Bermingham, Baker and McDonald (1979). This presentation is generally regarded as marking the launch of geodemographics in the UK.

Webber's classification was taken up by market analysis company CACI and was renamed Acorn (A Classification of Residential Neighbourhoods). Until the early 1980s this was the only commercially available neighbourhood classification in the UK.

The 1980s

In 1983 CACI introduced a new version of Acorn using the 1981 census. Throughout the remainder of the 1980s, a number of competing discriminators were launched by new entrants into the market, including PiN (Pinpoint Identified Neighbourhoods) from Pinpoint Analysis and Mosaic from CCN Systems (now Experian). In 1987 Pinpoint launched FiNPiN (Financial PiN), which was designed for use in financial services and so was the first market-specific system. Most of the classifications in the 1980s made use of census enumeration districts (EDs) to define their neighbourhood geography. The ED was the smallest unit of geography for which census results were output – an ED in England and Wales contained around 180 households, on average. EDs had been designed by the census offices for data collection purposes and were not ideal as an output geography.

The 1990s

Following the 1991 census, a host of new and updated geodemographic systems appeared, often making use of lifestyle survey data collected from large-scale direct mail surveys. During the 1990s we saw a move towards 'one-to-one' marketing, which led to developments of individual-level discriminators and 'fusions' between market research and customer data. An example of the latter was the FRuitS segmentation for consumer financial services, which assigned each adult individual to a financial segment (see Leventhal, 1997).

The 2000s

A new generation of classifications was launched from 2004 onwards, following the release of output from the 2001 census. One of the most significant advances in this census was the creation of a purpose-built small-area geography for census output, replacing the EDs used in previous decades. The new output area (OA) geography was based on computer-generated combinations of postcodes designed to create consistent areas, each containing around 125 households.

The census access development brought free access for most census output, and therefore stimulated a larger number of users to work with the data. Possibly as a consequence, the portfolio of geodemographic classifications widened, and included a number of smaller entrants into the market. Later in the decade, ONS also became a classification supplier, by launching the Output Area Classification (OAC), which was developed in partnership with the University of Leeds.

The 2010s

This decade has seen the growth of open data – sources of public information that are made available to users free of charge (or at low cost), under specified terms and conditions. The tabular output from the 2011 census is the largest open dataset to be released. Small-area statistics have again been produced down to OA level, making use of the geography created for 2001. By mid-decade, eight classification systems have been launched, including a new version of OAC from the ONS in partnership with University College London. These current products will be described more fully in Chapter 4.

The 2020s

Although this part of the history has yet to happen, the ONS Beyond 2011 programme has carried out a thorough review and consulted with users on options for census taking in the future. The programme concluded that a primarily online census should be conducted in 2021 and combined with greater use of administrative data sources, and the UK government has agreed with this recommendation. Therefore, the intention to carry out a 2021 census appears to be definite, which should underpin geodemographics in the UK for the foreseeable future. Chapter 11 discusses these developments and looks to the future.

Other methods of market segmentation

Market segmentation is an approach that entails dividing a marketplace into groups of consumers with common needs, and then creating and implementing different strategies for marketing to them. Geodemographic segmentation sits alongside a number of other techniques for creating these groups. Commonly used methods include:

- standard demographics such as age, sex, social grade and region – the traditional controls employed in market research;
- demographics that discriminate best between buyers and non-buyers of a market – the choice of discriminator tends to depend on the nature of the market; for example, life stage is important in financial services, where financial products are geared to factors such as age, marital status, employment and retirement;
- attitudes and psychographics – the attitudes and values held by consumers;
- channel – the method by which the consumer communicates with the supplier to make purchases, eg in store, by telephone, by post or online.

There is no single 'best' method of market segmentation – for any business use, the choice should depend upon a number of factors, including:

- first and foremost, the primary purpose of the segmentation and how it will be used;
- characteristics of the market in question;
- the factors or attributes that drive consumers to purchase in that market;
- the importance of the various channels used for communicating with consumers and for making purchases, eg online, offline (in store, by telephone or post);
- the importance of knowing the geographical locations of consumers in each market segment.

All methods of market segmentation will correlate or have relationships, to some degree, with geodemographic discriminators. In the following paragraphs, we pick out some of the widely used segmentation criteria that are likely to be of interest, and we comment on their relationships with geodemographics.

Social grade

Developed in the 1950s, the Institute of Practitioners in Advertising (IPA) social grade is the main socio-economic classification used in advertising, marketing and in market research. Social grade is employed for many purposes, such as defining target markets, and as a 'currency' for buying and selling advertising. Six grades (A, B, C1, C2, D and E) cover the range of occupational groups and also those who are retired, unemployed or not working for other reasons. The grade assignments for different occupations, together with the rules for grading those who are not working, are maintained by the Market Research Society (MRS) (see MRS, 2010).

Doubts are sometimes raised about the validity of capturing and coding social grade in research surveys, and alternative classifiers have been devised from time to time (see, for example, O'Brien and Ford, 1988). However, social grade has outlasted all other methods, largely because it discriminates well for a wide range of products and services, including newspaper readership, TV viewing and product purchasing. At the same time, social grade is heavily ingrained in marketing and advertising processes.

The UK Census has never measured social grade, although it does capture many of the characteristics that assist in grading people, including occupation, employment status, working status and qualifications. For 2001, members of the MRS Census and Geodemographics Group developed a model for deriving an approximate social grade for each individual and household, from demographics collected in the UK Census. The census offices applied the algorithm in order to create output tables on approximate social grade (see Meier and Moy, 2004). The model was rebuilt for the 2011 census and applied again, generating a wider range of tables; details of the development are given by Lambert and Moy (2013). An example illustrating the usefulness of census social grade is included in Chapter 5.

The demographics used to assign social grades are generally employed when building geodemographic classifiers, therefore it is hardly surprising that social grade and geodemographics tend to be strongly associated with one another. Users who require their locational segments to mirror their advertising targeting are likely to select social grade as their discriminator; others seeking locational segments based on a wider range of characteristics are more likely to opt for geodemographics.

Life stage

In the late 1980s, O'Brien and Ford introduced an alternative discriminator termed life stage, which overcame the technical problems in capturing social grade. Life stage was created from a combination of attributes such as age, marital and working status and presence of children, the outcome being a set of life-stage groups. Although there is no standard definition of life stage, the concept has been recognized as an important discriminator in many markets. Two life-stage classifications are available from the 2011 UK Census: 1) life stages of adults aged 16+ based on age, presence of children, age of youngest child and size of household; 2) life stages of household reference persons (the closest census equivalent to chief income earners) based on age, presence of children and size of household.

As for social grade, life stage can be expected to have a strong relationship with geodemographics and any choice between them should depend upon the importance of life stage to the market under consideration.

Attitudes and psychographics

Attitude analysis focuses on the beliefs, motivations or values that people hold, while psychographics encompasses attitudes, lifestyles and behaviours. These methods of segmentation are mainly used by advertising agencies in order to understand the motivations of consumers. Three examples of classifications in this space are:

- VALS (Values, Attitudes and Lifestyles): from Strategic Business Insights (originally developed by the Stanford Research Institute), VALS segments US adults into eight types using a set of psychological traits and key demographics that drive consumer behaviour.

- Social value groups: a classification of people based on a survey of their outlook and their social values. The classification draws heavily on Maslow's hierarchy of needs (Maslow, 1943, 1954).

- 4Cs: Young and Rubicam's 4Cs classification (cross-cultural consumer characteristics), which is a tool for identifying types of individuals based on goals, motivations and values.

Although these types of segmentations can yield a greater understanding of the consumer, their relationship with geodemographics tends to be weaker. So it is more difficult to locate members of an attitudinal segment 'on the ground'. Therefore, these techniques are best used alongside geodemographics, rather than making an 'either/or' choice.

Ethnic origins

Ethnic and cultural origins can determine the types of products that people purchase, their choices of media such as newspapers and radio stations, and other aspects of consumer behaviour. One segmentation system designed to provide this information is Origins, from OriginsInfo. Origins classifies people based on first and last names, and identifies the part of the world from which their ancestors are likely to have originated.

The Origins segmentation is not limited to the UK – its scope is international. It provides complementary insights and so is best used alongside geodemographics.

Can geodemographics actually predict purchasing behaviour?

The Luton experiment

Geodemographic classifications have been used in the UK since the late 1970s, and have contributed to countless site location models and targeting projects, without any formal proof that they can actually predict how people living in a neighbourhood behave. In 1995, the MRS Census Interest Group formed a working party to examine this question.[1] The team included representatives from the Group Market Research Department at Whitbread, market research agency BMRB International and two market analysis companies, CACI and Experian.

After initially examining profiles of product users with the help of analyses from the Target Group Index (TGI), the working party decided that it should test whether a geodemographic classifier could actually predict differences in consumption at a neighbourhood level. It therefore set out to survey the consumption rates for neighbourhoods within a town, and compare the results with geodemographic predictions.

The test was carried out in Luton, and involved carrying out a product consumption survey in 18 neighbourhoods, sampled within three Acorn types. Using the survey results, the 18 neighbourhoods were clustered into three groups based on their consumption patterns. Each group was found to correspond to a particular Acorn type, and 15 out of the 18 neighbourhoods were in perfect agreement. The other three neighbourhoods were investigated further and the differences were explained.

A number of cases examining whether geodemographics actually 'works' – including the Luton experiment in greater detail – are presented by Harris, Sleight and Webber (2005). All of their examples demonstrate that geodemographics does indeed 'work' for its users.

Conclusion

In this chapter we have overviewed the principles of geodemographics, the analysis of people by where they live, and reviewed a simple worked example. We have seen that geodemographics had its historical origins in the late 19th century, was reinvented for the modern age in the late 20th century and should have a healthy future for the foreseeable 21st century.

We have identified the key strengths and weaknesses of the technique, and seen that it sits alongside other demographic and psychographic methods of market segmentation.

Note

1 The MRS Census Interest Group was founded in 1989 and became the MRS Census and Geodemographics Group (CGG) in the mid-1990s. The CGG is an advisory board of MRS. Details of its activities may be found at: https://www.mrs.org.uk/mrs/census_and_geodemographics_group.

BIG DATA AND GEODEMOGRAPHICS

02

Introduction

We are now living in the era of big data, where vast quantities of data are continuously being captured about consumers from online systems, in-car telematics, smart meters and so on. Geodemographics may appear to be redundant when so much customer-specific information is available.

The main aims of this chapter are:

- To discuss the trend towards big data.
- To explain why geodemographics is still relevant in the big data era, with the help of a section contributed by Richard Webber.
- To suggest a couple of ways for big data and geodemographics to coexist and complement one another.

The trend towards big data

There is no 'industry standard' way to define big data; however, a good working definition is offered by the McKinsey Global Institute (2011): 'Big data refers to datasets whose size is beyond the ability of typical database software tools to capture, store, manage and analyse.'

As Bill Franks (2012) observes, this definition implies that, as technology advances, the data that qualifies as being 'big' will change. So transaction databases that store, for example, supermarkets' electronic point of sale (EPOS) data and mobile-phone call detail records are huge in size and can provide detailed information about customers, but are technically no longer 'big data'.

Ironically, going back to the mid-1970s to the early days of geodemographics, the set of small-area statistics from the UK Census would have been viewed as being 'big' – if an equivalent definition had existed at the time.

Franks comments that true 'big data' is not just about the volume of records, it is also about velocity (the rate at which data is generated), complexity and variety. 'Complexity' refers to unstructured data, such as web logs, which need to be read and interpreted in order to extract useful information. 'Variety' implies handling a wider range of sources than existed in the past. For example, online behavioural data includes:

- searching for products and services using a search engine;
- browsing items online, viewing pages and videos;
- adding items to a shopping basket and completing a purchase (or abandoning the basket);
- posting product reviews and comments.

Unarguably, harnessing big data is a significant trend – the growth rate is such that if we quoted any statistics on it they would be out of date by the time this book is published. However, does this mean that sources of 'small' data are no longer relevant? The following section, contributed by Richard Webber, explains why, in his view, geodemographics is still relevant to marketers.

Why geodemographics is relevant in an age of big data

Richard Webber, *OriginsInfo.Ltd*

The idea of geodemographic classification was introduced to the marketing industry in the late 1970s. By comparison with other marketing innovations of that time its use has been remarkably resilient, notwithstanding the changes in the structure of the marketing industry, new media and the plethora of new sources of data that have become available, particularly in the internet era. The purpose of this section is to explain the reason for this resilience.

During the late 1970s, when geodemographic classification gained its first adherents, it overcame two serious problems that had been troubling the industry. The first benefit was that it could provide information on what sorts of people were buyers of very particular products and services – and it could do this quickly, consistently, cheaply and without the need to add a question to a consumer survey and wait for a response. The second was that it enabled direct marketers to target new customer recruitment campaigns using direct mail or door-to-door distribution with the level of precision they have become accustomed to when using print media, radio and television.

Since the mid-1970s the mix of applications that rely on systems such as Mosaic and Acorn has evolved in response to changes in the focus of marketing practices, new communications channels and new information management technologies. Prior to the 1970s the focus of many large companies had shifted from mass marketing to what was described at the time as 'target marketing'. This followed naturally from the declining share of commercial television in the marketing mix, and improvements in the ability of computers to store, access and manipulate customer information.

As computers became able to store and process ever larger amounts of data, large organizations began to realize that the information they had used to manage customer accounts could beneficially be linked together in relational form so as to produce a more comprehensive and detailed picture of the purchasing history of each customer. Marketers in large organizations began to realize that this information could provide an invaluable resource for identifying existing customers to whom other products in the company's product portfolio could be cross-sold.

There are two reasons why, contrary to the prognostications of some industry observers, customer data did not cause a decline in the use of geodemographic data. The first of these is that, whilst customer relationship management systems provided an excellent resource for cross-selling and up-selling in sectors where customers transacted frequently but at low value (such as in banking or telecoms), there remained many other sectors such as automotive and large household items, where the tendency of customers to make infrequent but high-value purchases limited the volume of transactional information held about them.

Second, in markets such as credit cards, where a consumer is likely to hold an account with multiple providers at any one time, previous transactional information is necessarily restricted to that part of the customer's wallet that he or she awards to any one supplier. In the absence of geodemographic information it is nigh impossible for the marketer to distinguish on the one hand the customer who is faithful to a single supplier but makes little use of the product category and, on the other, the customer who is an extensive user of the product category but who undertakes the majority of his or her transactions with a competitor.

In this situation, whilst a customer relationship management (CRM) system is capable of describing the profile of an existing customer's spend, it is not able to place this in the context of the opportunity that the customer represents for additional business. A geodemographic classification is in a much better position to help gauge the size of the opportunity. However, direct marketing has never been the sole use to which marketers have put geodemographic classification.

The late 1970s, when Acorn was first introduced, was a period when many retail multiples were expanding their presence on Britain's high streets. In those days, when a multiple retailer operated a limited set of formats, there would be a requirement to establish how well the geodemographics of a potential new store catchment area matched the profile of what constituted the profile of the profitable customer.

As logistics improved, multiple retailers became able to support a much wider variety of retail formats, indeed even to apply the concept of mass customization to the process of deciding what to stock on a store-by-store basis. Where geodemographics had been used primarily to decide where to open a new store, increasingly it became used to decide what product lines should be stocked in the store and how much shelf space should be allocated to different product categories.

From the beginning, geodemographic classifications were used to exclude certain types of neighbourhood from particular product marketing campaigns. Door-to-door distributors, for example, quickly recognized that they could improve the effectiveness of their clients' spend by avoiding the promotion of conservatories to households who lived in neighbourhoods of social housing and the promotion of garden furniture to households who lived in high-rise flats.

What the more perceptive marketers began to realize was something less immediately obvious but potentially more powerful. That is, that the types of people with whom a consumer comes into contact on a daily basis have a significant impact on the decisions that the consumer makes, whether consciously through emulation – keeping up with the Joneses – or unconsciously, due to people's innate tendencies to adopt the prevailing opinions of the social groups to which they belong.

Geographers refer to these influences as neighbourhood effects. Essentially their argument is that although two postmen, for example, one in Bexhill-on-Sea, the other in Jarrow, may have identical demographic profiles, they are likely to espouse radically different social attitudes and to have very distinct consumer preferences because of the different profiles of the people they encounter at work and in their own neighbourhood.

The influence of where a person lives on their attitudes and behaviour was identified very early on in the commercial use of Acorn when Ken Baker, who added Acorn as a new field on the BMRB's Target Group Index, was able to demonstrate how, despite having very similar readerships in terms of social grade, the *Daily Telegraph* and the *Guardian* appealed to people who lived in significantly different types of residential neighbourhood. The *Daily Telegraph* was more likely to be read in neighbourhoods of uniformly white, middle-class and property-owning residents. The *Guardian*'s readers were more likely to be found in middle-class enclaves in ethnically diverse inner-metropolitan locations, many of whose residents worked in communications, entertainment or the public sector and where the proportions of women with a university degree, and especially in the arts, was particularly high.

The distinction drawn by geodemographic classifications between neighbourhoods dominated by conservative or liberal attitudes to social issues renders them particularly useful for the mainstream political parties' general election campaigns. For example, it can be shown that were electors in each social class to support Labour or the Conservatives in identical proportions irrespective of where they lived, the Conservatives would not have won a single seat in the general election of 1997. That they won as many as 197 seats was due to 'working-class' voters having a higher than average propensity to support the Tories in predominantly middle-class constituencies and for the middle classes to be more staunchly Conservative in neighbourhoods where they made up a majority than in neighbourhoods and constituencies where they were socially isolated.

Neighbourhood effects are equally evident in the analysis of pupil performance in key stage tests. Geodemographic analysis shows that pupils who live in low-status postcodes tend to perform much better in schools with high proportions of pupils living in predominantly middle-class neighbourhoods than they do in schools that draw their pupils predominantly from other working-class neighbourhoods.

To a large degree the effect of neighbourhood on social attitudes and on consumption reflects the impact on our lives of the people we encounter on a daily basis. But there is increasing evidence, particularly from the United States, that neighbourhood differences are increasingly exacerbated by people choosing to live in neighbourhoods where they believe neighbours will be sympathetic to their own values and lifestyles. This motivation is particularly important among those with pre-school and school-age children. Indeed it is quite remarkable how adept households are at using visual clues to interpret how easily they would fit in within any kind of neighbourhood. Were this not the case it would be far less common than it is for estate agents to display properties on the basis of photographs of their exteriors.

Nor should we overlook how much consumption patterns can be affected by access to and availability of products on a local basis. Once retailers learn to locate and merchandise their stores using geodemographic classifications, residents can find it increasingly difficult to purchase products other than those for which there is a strong demand from their neighbours.

In more recent years the advent of digital communications channels has hugely increased the amount of information that can be derived about an individual and that can be used to sharpen the targeting of communications. There are four principal respects in which information tends to be more specific:

- Information that is known about an individual tends to be far more timely. Whereas in the 1980s the decision as to whom to target with a promotion might be based on spend in the last few months, today the decision may be based on activity in the past few minutes.

- Likewise, whereas in the 1980s a company might have targeted a 'good' customer on the basis of total quarterly spend, or even spend on apparel, today it is normal to target customers who have bought specific items.

- Prior to the digital era and the availability of big data, data-driven marketing could only be based on previous purchase history and response/non-response to specific mailings. Today, segmentation can be based on what products consumers have looked at on the web and how long they have stayed on any particular web page.

- Finally, segmentation is becoming more specific in relation to customers' locations. It is now possible to target consumers who are known to have passed a particular location or even to target them when they are approaching particular locations.

Fabulous though these opportunities are, the more specific our targeting capabilities the easier it is to lose sight of the contextual data that can help to explain both the social influences to which consumers are subjected and their motivations and aspirations. If the highly specific and the contextual can be kept in balance, then it is much easier to identify the correct medium of communication as well as the message.

In my opinion, a compelling example of the importance of contextual information is the 2014 Clacton by-election campaign, which resulted in the return of the first UK Independence Party (UKIP) member to Parliament. The dominant geodemographic segment in the Clacton constituency is labelled 'bungalow retirement'. This demographic is dominated by people formerly employed in mid-level white-collar jobs, many of whom used to live in 1930s suburbs of large cities. On retirement, many of them have capitalized on the increased value of their houses and retired to modest seaside properties, typically in a street of uniform properties built by speculative developers for people on the cusp of retirement.

The attraction of these streets to ex-urban migrants is that they will be surrounded by people of similar attitudes to themselves – and perhaps similar to the people who were their neighbours when they first bought their urban properties 30 years previously, people who are likely to deplore the increase in social and ethnic diversity of what had once been staunchly conservative suburbs. These are likely to be people who share a sense of responsibility for their own welfare, who are reliable neighbours and regular in their habits. They tend to value personal contact when they visit neighbourhood shops, they do not hanker after foreign holidays, find it difficult to adapt to new technology, and above-average proportions tend not to have frequent contact with their grown-up children.

Whilst big data can clearly differentiate one household from another within this stereotype, being able to describe groups of individuals – many of whom share a broadly similar life history – would have enabled UKIP to identify Clacton as an easy seat to win. This would have told UKIP which streets in the constituency they should focus their communications on, what messages would resonate the most, and which postcodes in the constituency fitted this geodemographic segment.

It is this contextual capability that, in my opinion, explains why geodemographics continues to be relevant to consumer marketers in spite of the vast array of digital data to which they now have access.

How big data and geodemographics can work together

Geodemographics and big data should not been seen as an 'either/or' choice – in fact, they can and should complement one another. Two ways in which this can occur are given below.

Filtering a target group using geodemographics

Geodemographics provides a practical means of specifying the desired target audience for a product. The target segments may be defined as an additional filter, when online platforms serve up adverts to consumers; therefore an ad can be displayed to those consumers who have shown their interest in the product (from recent browsing behaviour) and also belong to its target market. This use of geodemographics is discussed further in Chapter 7.

The contextual information conveyed by the geodemographic segment may also be used to select the delivery method, style and tone of voice for each marketing communication.

Creating geodemographic discriminators from big data

Through analysis and aggregation of big data, new geodemographic variables may be created. For example, the Census 2022 project at Southampton University (see: **http://www.energy.soton.ac.uk/category/research/energy-behaviour/census-2022/**) has explored the feasibility of estimating small-area census-type statistics from smart-meter data – anticipating the widespread use of energy smart meters by the early 2020s. More generally, the 2021 UK Census programme, discussed in Chapter 11, is developing the use of administrative databases for producing population estimates and enhancing census outputs. Within the same programme, applications of big data are also being explored.

Conclusion

In this chapter we have argued that geodemographics is still highly relevant – even in today's world of big data. To summarize the main points:

- The increase in use of customer data for targeting, over recent decades, did not cause a decline in the use of geodemographics, for two reasons:
  - There are many infrequently purchased markets, with limited volumes of transaction data, which may be targeted using neighbourhood classifiers.
  - Geodemographics provides a more holistic view of the consumer, for example enabling their total spend to be estimated across all suppliers in a market.
- Geodemographics quantifies neighbourhood effects, which explains why products with similar demographic profiles will appeal to people living in different types of residential neighbourhoods.
- In more recent years, the digital data about an individual tends to be highly specific – in terms of timeliness, products recently browsed online, items recently purchased and the customer's location. However, this can lose sight of the contextual information that helps to explain social influences, as well as consumer motivations and aspirations.
- While big data can deliver highly specific information, geodemographics can provide the contextual influence – this is why geodemographics continues to be relevant to marketers of consumer products and services.
- There will continue to be opportunities for big data and geodemographics to work together – firstly for combining digital and contextual information when marketing to consumers, and secondly for building new geodemographic variables from consumer data.

KEY DATA SOURCES   03

Introduction

Geodemographics brings together disparate data sources about the population and where people live, and boils this down into a shorthand description – for example, as a classification code for each neighbourhood. The main purpose of this chapter is to identify the principal data sources that are employed in building and deploying geodemographic discriminators.

> **KEY POINT**
>
> There is an important distinction between sources that are required for building geodemographic discriminators and sources that discriminators are linked to. Therefore, this chapter is organized under these two main headings.

The build sources are brought together by the system developer for the purpose of assigning a classification code to every neighbourhood. Therefore, the build sources must cover the entire population and provide values for all neighbourhoods being segmented. Once the segments have been created they are attached to a suitable file of locations such as a postcode directory. At that point, the build sources are no longer required and so they may be invisible to the user – yet these sources will have defined how each area has been classified. In Chapter 4, we look at how classifications are created from these build sources.

The linked sources are a wider range of datasets that the discriminator may be appended to, for further analysis. These sources will differ from user to user, but they do not require complete population coverage – they could, for example, represent a sample customer file or a test region. There are many possible linked sources, which are discussed later in this chapter.

Sources required for building geodemographic discriminators

The build sources include inputs used to analyse and classify areas, along with files required in order to apply geodemographics to other datasets.

Input sources

Census of population

As highlighted in Chapter 1, the population census has always been the primary source for building geodemographic products, and each decennial release of small-area census data triggers a new generation of discriminators. The census is a vast information resource – too large to describe fully here – the interested reader is referred to the UK census offices (see Appendix A for further details) and also to the Geodemographics Knowledge Base (**www.geodemographics.org.uk**) for links to websites and articles on recent censuses. The section below summarizes some key features of the 2011 UK Census, related to its geodemographic applications.

The 2011 UK Census

Three censuses were conducted on 27 March 2011, covering the four countries of the UK – England and Wales, Scotland and Northern Ireland. These were carried out by the Office for National Statistics (ONS) for

England and Wales, National Records of Scotland (NRS) and the Northern Ireland Statistics and Research Agency (NISRA), respectively.

This 'devolution' of the census into three operations used to cause inconsistencies in the questions that were asked, the output tables and the available output geographies, making it difficult for users to obtain a consistent set of results across all of the UK. To overcome this problem, the three census offices made an agreement in 2005 that they would carry out their 2011 censuses on the same day and would work together on census outputs, in order to harmonize them where possible and produce consistent statistics across the whole of the UK (see ONS, NRS and NISRA, 2012).

The confidentiality of the census is paramount. By law, all census forms are kept confidential and individual census returns cannot be released for 100 years – after this time, the records are made available and used for genealogical research. Likewise, census returns cannot be passed to other government departments or organizations. Maintaining confidentiality is considered to be essential in obtaining support from the general public – although completion of the census is required by law, strong public support is essential in order to achieve a high participation rate.

For confidentiality reasons, census results are only released for geographical areas down to a certain minimum size, so as to remove the possibility of identifying individuals and disclosing information about them. The census offices take special measures, collectively known as disclosure control, in order to protect the confidentiality of the data. To this end, two complementary strategies were employed for the 2011 census – targeted record swapping and restricting the level of detail shown in tables for very small populations.

Targeted record swapping was applied when creating the final database used for production of census outputs. This entailed swapping a small proportion of households between areas, in order to add 'noise' to local area data while leaving results unaffected at national level. The swapping was 'targeted' in the sense that households with unique or rare characteristics were more likely to be swapped and their 'swaps' were selected to be broadly similar on basic characteristics such as household size.

The second strategy – restriction of detail in tables on small populations – was applied at the table design stage, and implies that tables produced for small areas contain fewer cells than those for higher levels of geography.

Going back to the data collection in spring 2011, a more consistent set of questions was asked across the UK for households and individuals –

the core set of question topics is shown in Table 3.1. As can be seen, the questions paint a rich and multifaceted picture of the population living in every area in the UK – a valuable start point for building geodemographic discriminators.

TABLE 3.1 2011 census question topics (England and Wales)

Household Questions

- Address
- Names, number and types of residents
- Number and types of visitors
- Relationships between residents
- Visitors: name, sex, date of birth and usual address
- Accommodation type
- Whether self-contained
- Number of rooms
- Number of bedrooms
- Type of central heating
- Tenure
- Type of landlord
- Number of cars

Individual Questions

- Name, sex, date of birth
- Marital/civil partnership status
- Other address for 30+ days a year
- Type of other address (armed forces, work, student home, student term time, another parent's home, holiday home, other)
- Whether student
- Where live in term time (here, other address as above, another address)
- Country of birth (England, Wales, Scotland, N Ireland, Republic of Ireland, Other – write in)
- If not UK, when last arrived here
- If arrived in last year, how long intend to stay
- How is your health in general?

TABLE 3.1 *cont'd*

- Hours spent caring for others
- Daily activities limited by health or disability?
- National identity
- Ethnic group (White, Mixed, Asian, Black, Other)
- Main language
- How well can you speak English?
- Religion (None, Christian, Buddhist, Hindu, Jewish, Muslim, Sikh, Other – write in)
- Usual address one year ago
- Passports held
- Qualifications held
- Work activity last week
- If not working last week:
  - Whether actively looking for work in last 4 weeks
  - Whether available for work within 2 weeks
  - Whether waiting to start a new job
  - Non-work activity last week (retired, student, looking after family/home, sick/disabled)
  - Year last worked, if any

FOR MAIN JOB or LAST MAIN JOB:

- Type of job (employee, self-employed – with or without employees)
- Job title
- Job description
- Whether supervise others
- Main activity of employer
- Name of employer

FOR JOB LAST WEEK:

- Address of workplace
- How usually travel to work (work at home, underground, train, bus, taxi, motorbike, car – driver or passenger, bike, walk, other)
- Hours usually worked (<=15, 16–30, 31–48, 49+)

SOURCE: Martin van Staveren (presentation to MRS/CGG Census Seminar, July 2011)

The main data collection vehicle was a census form that captured characteristics of each household and its individual members and, for completeness, also recorded visitors. A separate type of questionnaire was used for those living in communal establishments. Questionnaires were delivered by post, and were designed for self-completion and postal return. An innovation for 2011 was that the census could also be answered online – although this option was not strongly promoted, around 16 per cent of completed questionnaires came via the online channel. This choice of channels, in combination with a more targeted follow-up of non-responding households, worked well to optimize response rates. Overall, ONS achieved a response rate of 94 per cent for the census in England and Wales, with at least 80 per cent response in all local authorities. This level of participation can be viewed as highly successful, and 2011 is widely regarded as the most successful UK Census in modern times.

The other key component of the UK Census, of crucial importance to geodemographics, is the geography used for small-area output. Prior to 2001, the smallest geographical unit for census output was the enumeration district (ED), which was also the unit of geography used for census fieldwork. Although EDs worked well as fieldwork units, they had drawbacks as output units – they varied in population size and they bore no relation to the postal geography system employed by most users. For example, ED boundaries often cut through postcodes, making it difficult to assign a customer postcode to the correct ED – and therefore to the correct geodemographic segment.

Therefore, at the request of users, the output geography for the 2001 UK Census was separated from the data collection geography. A new geography of output areas (OAs) was created for England, Wales and Northern Ireland, by combining adjacent postcodes based on their census results. These OAs were designed to have similar population sizes and be homogeneous in terms of housing tenure and dwelling type. The target OA size was 125 households, with a minimum size of 40 households and 100 people (see p 158 for further details). In Scotland, the target OA size was 50 households, as smaller output areas had been introduced in Scotland for the 1981 census; the minimum size was 20 households and 50 people.

For 2011, the census offices decided to reuse the OAs that had been created for 2001, in order to maintain stability and enable analysis of changes within small areas between the two censuses. Criteria were agreed for when an OA had to be redefined; for example, if the OA population had been doubled by new homes built between the two censuses, then the original OA was split into two new OAs. Overall, across England and Wales, 97.4 per cent of OAs were retained, while 2.6 per cent were changed.

As the smallest units available, the outputs for OAs are generally the most useful for geodemographic analysis. However, the census also uses OAs as the building bricks to create a wide range of other geographies. A greater level of detail can be provided on tables for higher-level areas, such as local authority districts, and so it is useful to be aware of these other geographies. These vary between countries of the UK – for example, for England, areas created from OAs include:

- lower- and middle-layer super-output areas;
- wards, parishes and local authority districts;
- counties and regions;
- postal sectors;
- parliamentary constituencies;
- workplace zones, a new geography for producing statistics on workplace populations.

A variety of outputs is available from the census, at different levels of geography. The main products include:

- key statistics and quick statistics: profiles of census variables, down to OA level;
- local characteristics: multi-way tables on two or more variables, down to OA level;
- detailed characteristics: more detailed multi-way tables on larger geographical areas.

The census also provides more complex products:

- origin-destination statistics that provide data on flows from one area to another: for example, travel-to-work flows and migration patterns;
- microdata – anonymized files containing coded census returns, selected at random from the census database.

Finally, a commissioned table service is available, enabling users to obtain non-standard analyses or tables falling outside the published product sets, subject to their being non-disclosive.

Open data

Output from the 2011 UK Census is a particular example of open data. According to Opendefinition.org, open data is 'data that can be freely used, reused and redistributed by anyone – subject only, at most, to the requirement to attribute and share alike'. There are many other open data datasets and new ones are continually being released. The impact of open data on geodemographics has been profound – the following contribution by Dr Robert Barr explains how this has come about. Bob Barr has worked with geographic information for more than 30 years and is a member of the Open Data User Group (ODUG).

Geodemographics and open data

Robert Barr

Geodemographics developed because of the wide availability of machine-readable census data following the 1966 mid-term census and the full census of 1971. Researchers such as Richard Webber were able to develop computer-based methods of classifying data for census output areas – at that time, this data was primarily available to universities and other researchers; universities also had the necessary computing facilities for such advanced work, on what by 1970s standards were large and complex datasets. While subject to Crown copyright restrictions, access to census data during that period was relatively open, allowing many researchers to experiment with alternative approaches to geodemographic classification.

However, following the change of government in 1979, the relatively open access to census data for researchers and government officials was questioned and Lord Rayner was commissioned by the government to produce a White Paper on Official Statistics, which was published in 1981. This recommended that 'information should not be collected primarily for publication (but) primarily because government needs it for its own business'. The 'Rayner Doctrine' that ensued ensured that those outside government paid a market price for government information; unfortunately this also led to fees being charged within government to those needing to use census data and a substantial reduction in the staff and the capabilities of the Central Statistical Office.

Census data, which had previously been effectively open, started to be available only through commercial census agents who had to pay a substantial annual fee to allow them to trade in the data. As these statistics are only available from the government department that is enabled under the Census Acts to conduct a census, there could be no open competitive market for census data. Only those companies able to pay the high entry price were able to trade in census data and all could take advantage of the 'monopoly rent' that was created by restricting access to a unique source of information.

As a result of these restrictions the ability to develop geodemographic classifications for general use was restricted to the high-value end of the private sector. Two dominant products emerged, Acorn from CACI and Mosaic from Experian. Both were developed using commercially available census output, later enhanced with proprietary data from other sources. Interestingly, Richard Webber, sometimes called 'the father of geodemographics', was involved in turn with developing both of these products.

Acorn and Mosaic, along with classifications from other suppliers, were strongly marketed and calibrated against a wide range of consumption patterns. Their predictive value was good enough for marketing purposes. Once the geodemographic groups with a tendency to consume an item, such as to purchase a certain product or read a particular newspaper, were identified then these pieces of information could be combined. Those wishing to market a product consumed by selected segments could advertise it in newspapers read by those groups, possibly in the regional editions where the consumption of the product by those groups was lower than elsewhere.

Because this geodemographically informed marketing method has become familiar and understood and has known results, managers in both the public and the private sectors have tended to continue using neighbourhood classifications. In the public sector the propensity of particular groups to consume particular services is of more interest than their buying behaviour. This allows services to be better targeted and provision matched to predicted demand.

The Crown copyright review of 1999 led to a much less restrictive approach to government statistical data. In particular the results of the 2001 UK Census became freely available for general use, effectively open data, because the government's right to charge a fee for the use of such Crown copyright material was waived under a series of licensing arrangements. Initially a simple waiver licence known as a 'click-use' licence allowed unrestricted commercial and non-commercial use of the data. Later an even less restrictive 'Open Government Licence' was introduced for such data.

However, in order to create a useful geodemographic discriminator, it is not sufficient to have census data alone. It is necessary to know what addresses, or at least what postcodes, fall in each output area; and, if the areas that have been classified are to be mapped, output area boundaries are required. The data used to match addresses or postcodes to census output areas was controlled by Gridlink, a consortium including Royal Mail, Ordnance Survey and ONS. Unfortunately, of these three organizations, only ONS had become committed to open data. Both Ordnance Survey and Royal Mail, operating as a trading fund and a government-owned company respectively, retained a business model based on selling data. As a result the essential look-up tables to match addresses to output areas remained only available commercially. So, while it was technically possible to create a free-to-use open geodemographic segmentation, the data required to use it was not openly available. Such a classification was produced as part of a PhD project at the University of Leeds and was known as the Output Area Classification (OAC).

Following the release of the 2001 census data, ONS also released output area boundaries as open data. These boundaries did not follow features or roads on Ordnance Survey (OS) maps but were derived from 'AddressPoint', an OS product that provides an OS grid reference for every postal address in the country. However, as output areas were nested within electoral ward boundaries, the ward boundary element of OS's BoundaryLine product was included in the ONS data. This led to an unusual challenge to the open data status of the output area boundaries.

OS had no objection to the boundaries being used for the very small output areas, but objected to output areas being grouped into electoral wards. This led to a complaint to the government Office of Public Sector Information (OPSI) from the Association of Census Distributors against OS. However, the complaint was not upheld because it was accepted by OPSI that OS's business model did not include providing open data and that a ward map derived from the ONS data would compete directly against some of the data in BoundaryLine.

The absence of essential map and address look-up files as open data prevented OAC from gaining traction among users. Most users in the private sector continued to use the commercial Acorn and Mosaic products as well as alternative commercially available classifications such as CAMEO and Censation.

It was not until about 2009, when in a high-profile announcement the prime minister, Gordon Brown, allowed some OS data, including BoundaryLine and Codepoint Open (a dataset that provides a map reference for every postcode), to be released as open data. At around the same time ONS agreed with the Gridlink consortium that, with a small number of fields removed, the ONS postcode directory giving a postcode to output area look-up could be released as open data. At this stage a fully open set of geodemographic tools – including OAC, census boundaries and look-up tables – was available, and interest in OAC increased.

However, the commercial products had not stood still. While OAC was based on 2001 census data and an active user group had formed, the commercial products had been enhanced and updated with more recent non-census sources.

The demand for these products remained high and the open OAC, despite being available without charge, was not a serious competitor to the familiar, well-documented and well-supported commercial systems.

Following the release of 2011 UK Census data in 2013, a new version of OAC was produced. This time, far more flexible tools and a more active user community are taking advantage of this open resource. A free look-up and mapping system have been developed by Chris Gale and Oliver O'Brien, and are available from and supported by the Consumer Data Research Centre at University College London, and the universities of Liverpool, Leeds and Oxford (see **http://public.cdrc.ac.uk/**). Open access to geodemographics is documented at **http://www.opengeodemographics.com/** and the OAC User Group have their own website at **https://plus.google.com/u/0/ communities/111157299976084744069.**

It remains to be seen whether the now well-supported and versatile open geodemographic tools dent the well-established commercial market for such data or complement it. The commercial companies are fleet of foot and are using open attribute data as well as proprietary sources, often from their own customers, to enhance their products. So, at the very least, the availability of an open competitor is spurring innovation in geodemographics.

This has been a specifically UK-based overview; however, open data and geodemographic classifications based on it are beginning to have an impact elsewhere in the world. Alex Singleton, a leading academic geodemographic researcher in the UK and a principal member of the OAC and CDRC teams, working with Seth Spielman from the University of Colorado, has produced a US geodemographic classification based almost entirely on non-census open data. In Singleton and Spielman (2014), they discuss the past, present and future of geodemographic research in the United States and the UK.

The prospects for flexible open, user supported, geodemographic classifications based on open data look extremely promising. However, for the foreseeable future these will coexist with the commercial offerings that have a firm place in the marketplace and are themselves evolving – both because of the competition and due to the increasing amount of available open data. It appears that the current open data revolution is going to stimulate innovation and the development of new, improved and more flexible geodemographic products. We have not seen this level of change since the birth of geodemographics in the 1970s and it promises to be very exciting.

Lists and lifestyle data

Although the UK Census provides us with the widest range of demographic characteristics for neighbourhoods, it was never designed to describe consumer behaviour such as what people purchase or the media that they consume. Also, it does not measure incomes or savings, nor can it capture lifestyles or interests.

One objective of the commercially available geodemographic discriminators is, however, to identify target segments for different types of markets such as consumer goods or financial services. To achieve this goal, geodemographic developers include variables from other sources that can help to predict behaviour, often derived from lists and databases such as property prices from the Land Registry or home values from council tax data.

Lifestyle data – databases of individuals compiled from commercial lifestyle surveys on people's behaviour and interests – originated in the United States and started to operate in the UK from the mid 1980s. Lifestyles were widely used for one-to-one marketing in the 1990s and reached saturation point, in terms of data collection, by the late 1990s. By that time, International Communications and Data (ICD) was carrying out biannual mailings of lifestyle questionnaires to all households on the electoral roll, supported by TV advertising. Computerized Marketing Technologies (CMT) distributed lifestyle questionnaires focusing more on fast-moving consumer goods (FMCG) – the questions were sponsored by grocery manufacturers and retailers. The third major player, National Demographics and Lifestyles (NDL) specialized in capturing lifestyle information via guarantee cards for manufacturers of consumer durables.

As a result of the huge volumes of lifestyle surveys captured by these companies, the majority of 'post 2001 census' geodemographic products included lifestyle data as inputs, along with census and other sources. However, by the early 2000s, lifestyle gathering was in decline and nowadays the information is captured mainly via more targeted surveys. Lifestyle data may still be employed as an input for building discriminators; however, the user would then be justified in questioning the recency of the lifestyle component.

It is more likely, nowadays, that lifestyle data would be employed for describing the various segments of a geodemographic classifier; for this purpose, a sample of recent lifestyle surveys could be used. The excellent data visualization tool available for Acxiom's Personicx product illustrates the usefulness of lifestyle profiling (see **http://www.personicx.co.uk/**).

Postcode Address File (PAF)

The Postcode Address File (PAF) is Royal Mail's definitive address database for the UK. PAF contains 29 million residential and business addresses to which mail can be delivered, together with their 1.8 million postcodes.

Some geodemographic products use data from PAF to derive input variables when building discriminators. For example, 'farm' addresses imply rural locations, while houses with names indicate larger properties.

Using the postcode of each address as the key, geodemographic discriminators may be appended onto PAF data and used for selective campaigns or for database building. Appendix B summarizes the structure of postcodes in the UK.

Geo-codes and boundary files

Geo-codes and boundary files are the key datasets that enable geodemographic information to be located geographically, such as for mapping and relating to the catchment area of a store or the readership area of a regional newspaper. Their existence and application may not be obvious to users who are solely employing geodemographics to provide segment codes; however, they are essential for geographical analysis.

Geo-coding implies enriching the description of a location, such as a postcode, with geographical coordinates (also known as grid references) so that it can be mapped or analysed geographically. For example, by geo-coding the postcode records for customers of a retail store, they may be mapped and examined in relation to the store's location. For a postcode, the geo-code would ideally be the centroid (ie centre of gravity) of the locations of its constituent addresses.

Boundary files are used for mapping larger geographical areas such as census output areas or postal sectors. Off-the-shelf boundary files exist for standard geographies, such as the administrative and postal geography systems, while bespoke boundaries (eg 15 minutes' drive time from a store) may be created using appropriate tools, such as geographic information systems (GIS).

Geographic information and analysis systems are able to store and manipulate geo-codes and boundary data – for example, defining the catchment area of a store (in terms of distance or drive time, say) and then identifying those customers who live within that catchment area. These tools are discussed in Chapter 6.

Sources that geodemographic discriminators are linked to

Market research data

By linking geodemographic classifiers to market research data, developers and users may obtain segment profiles for various subgroups or audiences identified in the research, such as brand buyers or newspaper readers. This enables classification developers to obtain deeper insights and richer descriptions for their segmentations, in order to convey this understanding to users.

For marketers who do not have access to data on their own customers, research profiles may be used to identify target segments for consumers in a given market and to estimate how its potential value varies across any geographical region. Research profiles are described and illustrated in Chapter 6, while estimation of market potential is discussed in Chapter 10.

Customer files

The ultimate test of a geodemographic classification is to append it to customer data and assess how well it discriminates for various subsets of customers, eg by products used, customer value or responders/non-responders to marketing campaigns. The mechanics of customer profiling are explained in Chapter 6, using a worked example.

Big data

Similarly, geodemographics may be applied to 'big data' sources such as website visitors or subsets displaying particular behaviour patterns, eg visitors who browsed certain products but abandoned their shopping baskets. The proviso, of course, is that the data source needs to be linkable to geography, implying that it includes postcodes or identifiable locations of visitors.

Conclusion

When considering or selecting any geodemographic classification it is important to understand how each of its data sources was used – for assigning neighbourhoods to the different segments or for helping to describe the segments.

While the UK Census is generally accepted to be the bedrock source for building geodemographic discriminators, the system suppliers frequently include additional inputs to measure different aspects of consumer behaviour.

The more widely that a segmentation can be profiled and interpreted using non-census sources, such as market research, customer files and big data, the deeper the level of understanding that will be achieved.

GEODEMOGRAPHIC CLASSIFICATION SYSTEMS

04

Introduction

For most users 'geodemographics' equates to 'neighbourhood classification' and this chapter focuses on these systems – how they are developed and which classifications are available in the UK market following on from the 2011 census.

The aims of this chapter are:

- To discuss what we mean by a neighbourhood classification and outline how such a system is typically built.

- To provide an overview of the general-purpose classifications that have been launched into the UK market, following the 2011 census.

- To explain the differences between general-purpose and market-specific discriminators, and provide examples of the latter.

- To give initial guidance on which product to choose, before reading the more detailed assessment strategy presented in Chapter 8.

What is a neighbourhood classification?

A neighbourhood classification assigns a type code to each small area, according to its characteristics – in other words, it assigns areas into groups that could receive different marketing treatments. There are obvious parallels with market or customer segmentations, except in this case the segments contain neighbourhoods, such as census output areas (OAs) or unit postcodes, rather than individuals.

A typical classification consists of around 50 type codes, which combine into about 10 high-level groups. The 50 codes may well be summarizing some 100 facts about each neighbourhood, drawn from the UK Census and other sources that were discussed in Chapter 3. Thus, the classifier reduces a large amount of information down to a single code, which is more manageable than the original 100 characteristics.

Once created, the segments are interpreted and given labels and descriptions, which enable users to understand and recall them. Many of the suppliers go beyond this, and provide analysis and visualization tools that help to picture the types, behaviours and lifestyles of their residents.

How neighbourhood classifications are built

As we will see in this section, the process of building a neighbourhood classification differs little from customer segmentation, when looking for the 'best' set of segments across many attributes – and just as much 'art' or expert judgement is required in creating and interpreting geodemographic segments.

Although the detailed methods differ from one classification to the next the process generally involves six main phases, which are discussed below. These correspond to the high-level stages in the Cross Industry

Standard Process (CRISP) for data mining, documented by a group of analytical software suppliers and clients (see Chapman *et al*, 2000). The same six phases apply to building any analytical model, although the detailed tasks will differ from project to project.

Phase 1: business understanding

The developers will start the process by setting some clear business objectives for the classification builder to follow. The development team may well carry out initial research into the geodemographics market and what users need, in order to gain a sufficient understanding of 'where to take' a new product before setting the business objectives. The business objectives will answer questions such as:

- What is the intended scope? For example, will the system be 'general purpose' – used across a wide range of sectors and applications – or will it focus on a particular market (eg financial services) or channel (eg online marketing)? These different approaches are discussed later in this chapter. For the explanation of phases presented here, a general-purpose classification will be assumed.

- Dependent partly on the scope, does the classification need to make use of certain data sources and where are they to be found? For example, will it be focused on 'wealth' or 'affluence'? If so, should it include data on house prices or council tax bands?

- Does the classification need to discriminate at the finest level of geography, ie unit postcodes, for commercial targeting purposes? If so, then much of the development will be undertaken at postcode level.

- Does the classification need to be updated over time, for changes in areas? If so, then it should make greater use of, or place higher weight on, continually updated sources.

- Or does it need to be an accurate segmentation of areas in the UK as at the 2011 census, with no future updates required?

Phase 2: data understanding

Given the wealth of information available from the UK Census and the other sources described in Chapter 3, the classification builder should not be 'short' of data – in fact, there will probably be an overload from some sources, but less that is available from others. The main role of the data-understanding phase is to evaluate each of the available sources and the attributes or variables within them, in order to select a candidate set of inputs. This serves several purposes:

- To make decisions about which sources should be included in the development and which to hold back for evaluating the segments, according to the business objectives that have been agreed for the classification.

- To identify candidate variables for each of these uses, identify any issues with the data and plan how to resolve them.

The data issues will include how to handle variables that are highly 'fragmented' – for example, census variables for attributes such as country of birth, ethnicity or religion may contain many small categories that do not warrant analysis in their own right, and so will need to be grouped or combined with other categories in some way.

Remember that the units of analysis for building a neighbourhood classification will typically be small areas such as census OAs or postcodes. Therefore, the initial source data on country of birth from the census, for example, will consist of counts for the numbers of people in each OA who were born in England, Scotland, Wales, Northern Ireland, Republic of Ireland and so on – rather than being a single 'country of birth' attribute value.

The first step in order to understand better how country of birth varies between neighbourhoods will be to convert those OA counts into percentages, based on the total number of residents in each OA. Having done that, the analyst will be able to examine the variation in the incidence of each country of birth, across OAs. Therefore some early data preparation (phase 2) is required for phase 1 – hence the process is somewhat iterative.

The analyst will be looking to decide the most appropriate levels of aggregation for the variables, so that they discriminate well between areas without being too sparsely populated.

Phase 3: data preparation

Before multivariate techniques such as clustering can be applied, the data inputs need to be converted into a single 'row and column' table, where each row represents a neighbourhood and the columns correspond to the variables included in the analysis. The construction of this 'data matrix' is the main purpose of the data preparation phase, which typically takes up the largest amount of time in any analytical project. The main outcome of this phase is the analytic dataset (ADS) for building the neighbourhood classification.

For inclusion in the data matrix, all of the sources will need to be expressed for the same neighbourhood units. For example, if the units are going to be census OAs, then any inputs at postcode level will need to be aggregated to OAs, before being merged into a single file at OA level. On the other hand, if the classification is to include census data and the units are to be postcodes, then the census variables will need to be estimated at postcode level.

Two technical issues that should be examined and resolved during the data preparation phase are **transformation** and **standardization** of the variables in the data matrix.

Transformation implies expressing a variable that is highly skewed in order to become more symmetrically spread, ie to come closer to a normal distribution. This property is desirable, especially from a statistical point of view, for carrying out significance testing. The use of transformations to correct for non-normality differs from analyst to analyst, depending on the perceived need for accurate statistical tests.

Standardization implies re-expressing each input variable, so that all variables are measured on the same scale. Without tackling the issue of standardization, characteristics with low incidence would differ less from area to area than those with high incidence; therefore, 'low incidence' variables would contribute less when clustering areas on similarities and dissimilarities across many inputs. Standardization is often achieved by transforming each variable into Z-scores (by calculating deviation from the mean, divided by standard deviation for each variable).

One further issue to consider is the **polarity** or **direction** of the variables. For example, suppose that the analyst is using the measures 'Aged 65 or over' and 'Retired working status'. These will obviously have higher values in neighbourhoods containing many elderly people – therefore, they work in the same direction and will be positively correlated. If, instead, the analyst planned to use 'Aged 65 or over' and 'Non-retired', then these variables would operate in opposite directions and so would be negatively correlated, which could be confusing at the very least! While this point might seem obvious and trivial on only two variables, it can be harder to detect when there are many inputs, as will be the case when building a geodemographic segmentation.

Handling correlated variables

A crucial decision is how to handle the issue of correlated inputs, which often arise – particularly when analysing groups of census variables. Continuing the previous example, 'Aged 65 or over', 'Retired working status' and 'Pensioner households' are all present in the census data – their penetrations at OA level will be highly correlated with one another because they are all measures of the same thing, which we might call 'old age'. If all three measures were included in the classification, then the importance of old age would be overstated – so the question is how to handle this issue.

Over the years, classification developers have employed two alternative approaches for dealing with correlated variables. The first approach is to identify this issue at the preliminary analysis stage, by analysing and inspecting the correlations between all variables. This can be aided by producing a minimum spanning tree, which is a graph that displays sets of variables with the largest correlations. Part of a tree is shown in Figure 4 – Figure 4.1a illustrates a partial set of correlations between census variables, which converts to the hypothetical minimum spanning tree in Figure 4.1b. For example, the high correlation between Students and University Education results in these variables becoming adjacent vertices in the minimum spanning tree.

FIGURE 4.1a   Example correlations between census variables

	Age 25–34	Age 35–44	Flat Dwellers	University Education
Age 16–18				
Students			X	X
Age 21–24	X			
Industry: Retail				
Industry: Education				X

NOTE: larger correlations are depicted by darker shading, smaller correlations by lighter shading (hypothetical example). Adjacent vertices in the minimum spanning tree are shown by white crosses.

FIGURE 4.1b Example minimum spanning tree (partial tree based on a set of correlations between census variables, hypothetical example)

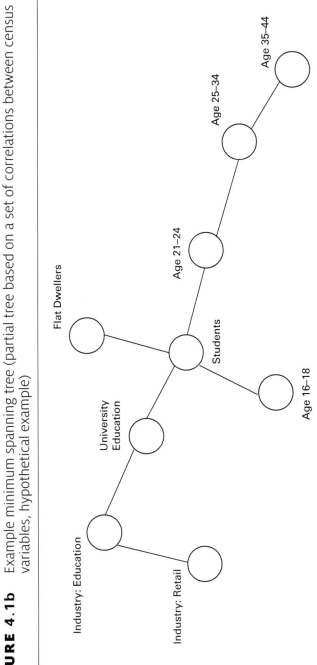

Having identified a correlated set, one variable is retained (typically that with the greatest number of strong relationships with other variables) and the others are excluded from the model development. Following this route, the developer obtains a smaller number of variables for building the classification; however, they are all 'key drivers' in discriminating between neighbourhoods.

The alternative approach for dealing with correlated variables is to employ a technique known as principal component analysis as the first stage in the classification development. This method, described in more detail below, essentially automates the task of identifying a set of correlated variables and replacing them by a single new component score, which in the above 'old age' example would be one new score replacing three correlated variables. Taking this route, a larger number of variables are included in the model development; however, they are first reduced down to a much smaller number of component scores.

There is no universal agreement on which of these two options is the better approach, but it is generally accepted that one or the other should be applied. Certainly, inputting a large number of variables, some of which are highly correlated, into a classification analysis would be an undesirable way to proceed.

Phase 4: modelling – building the neighbourhood classification

The modelling phase, to create the classification, is an extreme example of data reduction. It uses the candidate variables from the previous phase, and primarily applies the following multivariate analyses in combination:

- Principal component analysis (PCA) – optionally employed to reduce the inputs down to a smaller number of summary scores.
- Cluster analysis – to group similar neighbourhoods together into area segments.

Table 4.1 identifies and explains the multivariate analysis techniques that are most commonly employed for building neighbourhood classifications.

Whether or not to use PCA

PCA works well in geodemographics as a technique for 'extracting' the underlying themes or dimensions from a set of correlated inputs such as

TABLE 4.1 Techniques often used for building neighbourhood classifications

Principal Component Analysis (PCA)

A data-reduction technique that transforms a set of correlated variables into a smaller set of linear combinations that account for most of the information in the original set.

The aim of PCA is to obtain new variables (components) that explain as much of the variation in the original data as possible, with the smallest number of components possible.

Cluster Analysis

A family of techniques for grouping neighbourhoods based on the similarities between them.

The aim of cluster analysis is to create groups that display small variation within them, relative to the variation between groups.

K-Means Clustering

A technique that creates groups by partitioning the data on neighbourhoods into subsets and reallocating the assignments until an acceptable set of groups is obtained.

The aim is to achieve a solution where the neighbourhoods within a group are nearer to one another than to the neighbourhoods in any other group.

Hierarchical Agglomerative Clustering

A technique that successively joins together neigbourhoods into groups, by fusing together the two closest objects. The process continues in this way, fusing neighbourhoods or groups together, until eventually one single group is formed containing all neighbourhoods.

The aim is to create a hierarchy of clusters, from which one or more sets of high-level neighbourhood categories can be selected.

census variables. These dimensions can be converted into useful discriminators in their own right, and are discussed further in Chapter 5.

As suggested above, the use of PCA in constructing neighbourhood classifications is not universal. In the past, some developments have employed PCA – for example Charlton, Openshaw and Wymer (1985) report its use in the construction of SuperProfiles using data from the

1981 UK Census. From the author's personal knowledge, Pinpoint Analysis also used PCA, when building Pinpoint Identified Neighbourhoods (PiN) at around the same time. Pinpoint took a starting set of 104 census variables and employed PCA to extract 10 components from which PiN was created.

More recently, Harris, Sleight and Webber (2005) reported Experian's view – namely that the use of PCA can blur the fine differences between clusters, and that employing discriminatory raw variables can produce sharper results.

In the 1980s the geodemographics market was highly competitive and the use of PCA became a topic for debate between suppliers. Nowadays, it is accepted that this decision rests with the classification builder and that what is most important is that the input variables (or components) have been selected with care.

The use of cluster analysis for segmenting neighbourhoods

Some form of clustering is universally employed in order to segment neighbourhoods – based either on selected census variables or on principal components. The detailed approach again differs between systems; however, the use of k-means cluster analysis (outlined in Table 4.1) is most common. This technique is a highly efficient way of partitioning a large dataset; however, its main drawback is that the desired number of clusters has to be decided in advance. The classification builder will often proceed by producing and comparing a series of possible solutions and testing them in order to select the 'best' outcome or, in other words, the most useful set of groups.

Geodemographic classifications are usually hierarchical, with two or three levels in the hierarchy. This helps to widen their use – the top-level categories may be employed more strategically, while the low-level segments can be tactically applied to locate specific target types. To construct a classification, the analyst may take a 'bottom up' approach in which the low-level segments are first developed, using k-means analysis. Alternative clustering runs will be tested, with differing numbers of segments, in order to obtain an acceptable solution. The final segments are then employed as 'building bricks' for an agglomerative analysis, which creates a hierarchy of clusters – from which a most useful top-level set of categories can be derived.

Alternatively, the builder may decide to adopt a 'top down' segmentation approach, in which the top-level categories are identified first, by applying k-means analysis across all neighbourhoods. A number of cluster solutions will be created and examined, containing different numbers of categories in order to arrive at an acceptable high-level classification. Then, for each of the resultant high-level groups, its constituent neighbourhoods are extracted and a further clustering is carried out to identify finer segments, thus forming the low-level classification.

An example of a current 'top down' neighbourhood classification is the 2011 Output Area Classification (OAC), which was developed by ONS in partnership with researchers at University College London. The current product followed on from a highly successful 2001 OAC, which was also developed 'top down', and this approach was adopted again for 2011. The methods employed in building 2011 OAC are clearly documented – see ONS (2014) – and are summarized in Figure 4.2. This overview shows that OAC was based on 60 census variables and did not employ principal components analysis for data reduction.

Labelling the segments

The final step in building the classification is to interpret and describe each of the segments. If the earlier steps within this phase can be described as 'science', then labelling is definitely an 'art'. It entails profiling each segment by all the available inputs and other sources, such as market research and lifestyle databases, and examining these profiles. The interpretation process is not dissimilar to being a detective – searching for clues about a segment, joining these together to form a valid description and capturing that in a 'pen portrait' and a label that communicates the key differentiators.

Phase 5: evaluation – describing and testing the classification

The main aim of this evaluation is to assess the usefulness of the classification and whether it achieves the business objectives that were set for it in phase 1. The methods most frequently applied include mapping and sense checking the locations of the neighbourhood types, and validating the discriminator against other datasets – typically using customer files and market research databases.

FIGURE 4.2 Methodology overview for 2011 Output Area Classification

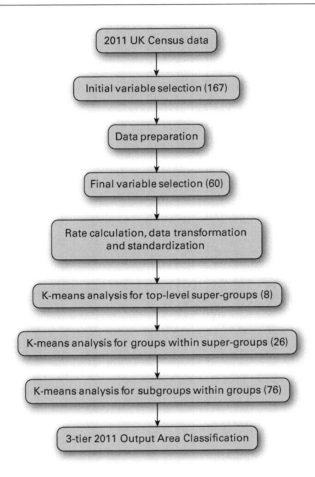

SOURCE: Methodology Note for the 2011 Area Classification for Output Areas, Office for National Statistics. Adapted from data from the Office for National Statistics licensed under the Open Government Licence v.3.0.

The validation stage is essential in order to decide how well the new classification would perform, if it were deployed amongst its target user base. This would include making comparisons against existing discriminators, such as any previous version of the same product.

Phase 6: deployment

The deployment process involves packaging the classification for use by internal and external systems. This includes production of a postcode directory that attaches the classifier's codes onto every unit postcode, and developing a range of materials and documentation for users. Supporting materials will include graphical summaries to help visualize the pen-portrait descriptions and to illustrate segment profiles for a variety of target audiences.

Current neighbourhood classifications in the UK

By the end of 2014, eight 'post 2011 census' neighbourhood classifications had been launched into the UK market. Although they are all general-purpose UK classifiers, each product differs from the next in a host of ways – including the small areas that are classified, the data sources employed, the number of tiers in the classification and the number of types identified at each tier. These features are summarized in Table 4.2, together with a web link for each classification or supplier. The corresponding table is also available online, updated for more recent changes in the market (see either **www.koganpage.com/product/geodemographics-for-marketers-9780749473822** or **www.barryanalytics.com/geodems4marketers**).

Each of the products in Table 4.2 takes a slightly different 'angle' to position itself uniquely in the geodemographics marketplace. The following paragraphs summarize their key points.

Acorn

When constructing Acorn, CACI employed a wide range of datasets, including public-sector open data and administrative data, as well as lifestyle data and proprietary databases built in-house. Although census data was used in the classification, the contribution from this source is being progressively down-weighted. This approach means that CACI will be able to refresh the data inputs and keep Acorn up to date over the coming years.

TABLE 4.2 Current geodemographic classifications in the UK

Classification System	Supplier	Geographical Units Classified	Number of Clusters	Information Sources Used
Acorn	CACI (see: **http://www.acorn.caci.co.uk/**)	postcodes	6 categories, 18 groups, 62 types	Land Registry, National Register of Social Housing, DWP benefits, UK Census, lifestyles, administrative and commercial datasets
CAMEO	Callcredit Information Group (see: **https://www.cameodynamic.com/**)	postcodes	10 groups, 68 categories	UK Census, open data, Define, YouGov, TGI, credit risk
Censation	AFD Software (see: **http://www.afd.co.uk/**)	postcodes	52 types	UK Census, British Population Survey, Land Registry
Mosaic	Experian (see: **http://www.SegmentationPortal.com**)	postcodes	15 groups, 66 types	Experian ConsumerView, UK Census, accessibility, open data, education

TABLE 4.2 *cont'd*

Classification System	Supplier	Geographical Units Classified	Number of Clusters	Information Sources Used
Output Area Classification (OAC)	ONS/University College London (see: **http://www.opendataprofiler.com/2011OAC.aspx**)	output areas	8 super-groups, 26 groups, 76 subgroups	UK Census
P² People and Places	Beacon Dodsworth (see: **http://www.beacon-dodsworth.co.uk/**)	output areas	16 trees, 44 branches, 180 leaves	Census, British Population Survey
Personicx	Acxiom (see: **http://www.personicx.co.uk/**)	postcodes	55 types	Customer data, consumer surveys, food and expenditure surveys, credit and expenditure surveys, house prices
Sonar	TRAC Consultancy (see: **http://www.tracconsultancy.co.uk/**)	output areas	6 life-stage groups, 4 wealth quartiles, 80 codes	UK Census, council tax bands, Land Registry, wealth, consumer activity

NOTE: for the latest version of this table, see: www.koganpage.com/product/geodemographics-for-marketers-9780749473822 or www.barryanalytics.com/geodems4marketers

CAMEO

CAMEO was built using a combination of sources including the UK Census, other open data, TGI and YouGov surveys. In addition, some of Callcredit's in-house databases were employed – Define, providing lifestyle survey and transactional data, as well as in-house credit-risk systems. CAMEO is updated monthly from these sources, so as to reflect changing consumer characteristics and people's movements between segments, as their lifestyles or prosperity change. In addition to the postcode classification, CAMEO is also available at individual and household level.

Censation

Censation uses data from the 2011 UK Census, the British Population Survey and the Land Registry. Unusually for a commercial product, Censation is provided free of charge to customers of AFD Software Ltd who use their address management solutions.

Mosaic

The latest version of Mosaic was built using Experian's ConsumerView database of the UK adult population. Additional sources, including census, retail accessibility, education and other open data were also employed. A Mosaic segmentation at household level was developed first, and then other products followed. As a result, postcodes, households and individuals are assigned to a common structure of Mosaic segments.

Output Area Classification (OAC)

The 2011 Output Area Classification (OAC) was produced by University College London in collaboration with ONS. OAC is a 'traditional' classification, using only census data, and is unusual in being open data, ie free to use. The open methodology employed in OAC means that it may also be used as a tool for building bespoke classifications.

P^2 People and Places

The new version of P^2 includes an economic aspect to the classification. This has been achieved by defining nine study regions across the UK, based on economic variables. The approach allows P^2 to take account of differences between regions and also update the region descriptions over time. A base set of clusters was created within each of these regions, then these clusters were further grouped together to build the main classification.

Personicx

The primary input into Personicx was Acxiom's lifestyle survey database – no census data was used. The solution operates at all levels of granularity – postcode, household and individual, and forms part of Acxiom's global Audience Operating System. The segmentation can be visualized online via the Personicx 'Eye'.

Sonar

TRAC Consultancy built Sonar on output areas, after finding that there was little improvement to be gained by classifying at full unit postcode level. Sonar uses census data in combination with public-sector sources, and is post-stratified into groups by life stage and affluence.

Market-specific classifications

So far, we have focused on general-purpose or 'vanilla' classifications – as far as most users are concerned, geodemographics consists only of these products. The general-purpose classifiers are the 'flagship' segmentation systems promoted by the various system suppliers and are designed for use across all consumer markets in the UK, as well as for all distribution channels and business applications.

Alongside the 'vanilla' products, there is a variety of market-specific classifications that focus either on a specific country or region within the UK, or are designed for a particular industry sector or aspect of consumer behaviour. There are two potential advantages of taking a 'market-specific segmentation' approach: first, by only using relevant data inputs for that market, the segmentation should discriminate better than the general-purpose systems. For example, a classification that focuses on one region, such as Scotland, may identify characteristics or nuances of Scotland's population that are less obvious within a UK analysis. Or a segmentation designed for a specific sector, such as financial services, may be constructed using discriminatory inputs for that sector – this can include the use of market research variables to guide the formation of high-level segments from low-level geodemographic clusters. Second, the segmentation will have been interpreted and described using language that is relevant for that market, and so should be more actionable for users.

Table 4.3 provides some examples of market-specific classifications, currently available in the UK, from each of the main suppliers that specialize in these more bespoke products.

TABLE 4.3   Examples of market-specific small-area classifications

Supplier	Market-Specific Product	Description
CACI	Wellbeing Acorn	Segmentation of the UK on health and well-being
	Social Scene Acorn	Segmentation for the leisure, eating-out and drinks market
Callcredit Information Group	CAMEO Income	Postcode classification for assessing levels of household income
	CAMEO Financial	Postcode classification for assessing credit risk
	CAMEO Investor	Postcode classification for assessing financial sophistication and wealth through shareholdings
	CAMEO Unemployment	Postcode classification for assessing economic inactivity
	CAMEO Property	Postcode classification for assessing wealth through council tax bands and property prices
	CAMEO Welfare	Postcode classification on levels of deprivation
	CAMEO Scotland	Country-specific segmentation for Scotland
Experian	Mosaic Public Sector	Designed for use by public services organizations
	Mosaic Scotland	Country-specific segmentation for Scotland
Future Cities Catapult	Whereabouts London	Region-specific segmentation for London

Customized classifications

Several of the geodemographics suppliers have created sets of 'micro clusters' below the lowest tier of their main classification. These micro clusters can be aggregated to form new segmentations for specific markets, driven by market research, or bespoke systems for clients using customer data (see Chapter 10).

While the statistical accuracy and discrimination may not be as great for a classification built using off-the-shelf clusters, this route is highly effective in allowing relevant external sources to drive the creation of the segments. At the same time, the required development time and resources are greatly reduced by taking this approach.

How classifications are updated

Two types of maintenance and updating are required for any neighbourhood classification – the first is to reflect changes in characteristics of areas over time, while the second is to incorporate new or renumbered postcodes.

The ability to update for area change is dependent on the data sources that were used for building the classification. For those systems that are based solely on 2011 UK Census data, there is no way to reclassify areas that have changed – at least not until 2021 census results become available during the early 2020s. In practice, this limitation is generally not as bad as it sounds – in the majority of neighbourhoods, characteristics such as dwelling types and status alter very slowly over time. The changes only become significant in areas undergoing redevelopment, gentrification or decline, and in major housing developments.

This limitation is removed if a classification is built entirely using refreshable sources such as open data and transaction files. In this case, it is possible in principle to update the small-area data from time to time and reclassify areas. However, the majority of systems are based at least partly on census data, and therefore updating will continue to be problematic.

The second type of updating, for changes in postcodes, is more straightforward to handle and should be part of the standard maintenance process for each system. Each quarter, Royal Mail typically issues several thousand new postcodes – those that are small user codes, for new homes, can be identified and assigned to their appropriate neighbourhood segments. Updated classification directories are regularly produced and issued to clients for implementation.

Choosing a classification

In this chapter, we have identified eight general-purpose (Table 4.2) and 12 market-specific (Table 4.3) classifications currently available in the UK market. How should a new user choose which product is best, from such a wide range?

The analytical answer can be found in Chapter 8, which suggests an approach to evaluating discriminators by profiling samples of customer data. However, here are some preliminary thoughts to consider. First, for all of the commercial systems, by selecting a product the user is also deciding to work with its supplier. Therefore, it is advisable to consider which suppliers offer the best 'fit' with the user's wider needs and company culture, as this decision is likely to mark the start of a long-term relationship.

Second, the choice of a classification system should primarily depend upon its intended uses and how accurately it discriminates for these applications. Therefore, the user is best advised to identify uses and priorities first, before seeing how well each of the products would meet their needs.

> **KEY POINT**
>
> The user is well advised to ask any potential supplier to provide details of the variables that were actually employed in classifying neighbourhoods, how these variables were used and how much weight they carried in the solution. Nowadays this question is more challenging to answer, as classifications are developed in highly sophisticated ways, using different subsets of variables for the various parts of the build process. However, it is worthwhile trying to check whether variables that are known to be important for the user's market are helping to determine the segmentation. Overall, it is generally more important – in terms of performance – for a classification to be based on a carefully selected set of variables, or principal components, rather than to use a huge number of variables. So, beware of classifications built using all possible variables!

All suppliers should be willing to provide sample results, showing how their systems profile a sample customer file. These are a good way to get started, leading on to the more analytical evaluation described in Chapter 8.

Conclusion

This chapter has reviewed the way in which neighbourhood classifications are developed – there is no standard process or 'best practice' approach, because the outcome is geared to the aims and requirements of the classification builder.

We have seen that, in the UK alone, there are currently no less than eight general-purpose systems (Table 4.2) and 12 market-specific products (Table 4.3) – giving a great array of options for users. The choice of system supplier is as important as the choice of classification – the latter choice can be tested; the former, however, is down to assessment of 'fit' and 'company culture'.

OTHER TYPES OF GEODEMOGRAPHIC DISCRIMINATORS

05

Introduction

Moving beyond the ubiquitous neighbourhood classification takes us to a myriad of other discriminators and indicators, which form the main focus of this chapter.

The aims of this chapter are to explore these different types of products, including:

- Specific characteristics available from the UK Census – often known as census 'raw variables'.
- Indices and indicators that allow neighbourhoods to be scored and prioritized.

- Fuzzy classifications, which operate on a probabilistic basis.

- Individual-level discriminators that bring geodemographic segmentation down to individual household or person level.

Census raw variables

In some situations, consumer behaviour may depend heavily on a small number of demographics such as social grade, ethnicity or country of birth. Where those key characteristics were collected in the UK Census, then it can be highly effective to use those census outputs as discriminators, rather than attempting to achieve an equivalent result using a neighbourhood classification – because classifications are based on a wider range of measures and so will always dilute the effect of any specific variable.

The raw variables available from the 2011 UK Census were listed in Table 3.1, and further variables were derived at the processing stage, including approximate social grade (see Chapter 3). Census output may be obtained for a range of different geographies, including OAs, super OAs and postal sectors, enabling users to analyse the data at appropriate levels according to their business applications. The following contribution from Simon Whalley illustrates the usefulness of census raw variables, in the context of analysing results from the Scottish independence referendum held in 2014. Simon is data manager at census distributor Beacon Dodsworth, where he is responsible for analytics and segmentation development.

Insight from social grade – how social grade can help in understanding social trends such as calls for home rule

Simon Whalley, *Beacon Dodsworth Ltd*

Analysing and understanding geographical patterns are key to all types of organizations, whether they are within the public sector, academia or the commercial sector. Whatever the type of indicator used to try and understand a pattern, the key element is that it is discriminative. It highlights and explains differences within the dataset being examined. An indicator that is flat is no use at all as it tells you nothing. Intuition can be very useful, but it can be the case of trying out different indicators until one starts to explain the patterns that are shown. The rule of 80 per cent perspiration and 20 per cent inspiration is often true!

Here we use social grade to look at the results from the Scottish independence referendum held on 18 September 2014. We all know that the result was a 'No', but it is still interesting to look at the way that people voted and any potential patterns that emerge in order to try and understand why they voted the way they did. Figure 5.1 shows the percentage of people who voted 'Yes' for each local authority (LA) within Scotland. Only four LAs had a 'Yes' majority (shown in Table 5.1).

TABLE 5.1   Results – Scottish independence referendum 2014

Local Authority	Yes Vote (%)	No Vote (%)
Dundee City	57.35	42.65
West Dunbartonshire	53.96	46.04
Glasgow	53.49	46.51
North Lanarkshire	51.07	48.93

SOURCE: BBC at http://www.bbc.co.uk/news/events/scotland-decides/results

Looking at Scotland as a whole, Figure 5.1 shows how the local authorities around Glasgow, along with Dundee City, were more likely to vote 'Yes'. The Highlands and Western Isles had considerable support, although not a majority. The east of the country and the Borders were largely against independence. The pattern was almost an east/west split.

Social grade is a social classification that has been used for over 50 years by the advertising and marketing industries, but never available from the UK Census. Members of the Market Research Society's Census and Geodemographics Group developed an approximation to social grade for the 2001 census from census variables including occupation, employment status, qualification and tenure.[1] It has subsequently been updated for the 2011 census.[2] In each case, the census offices applied the algorithm to the population and produced output tables by approximate social grade. In this classification each household reference person (HRP) has been grouped into one of four categories based upon their occupation and employment status. These have been summarized in Table 5.2.

FIGURE 5.1   Scottish referendum results by local authorities

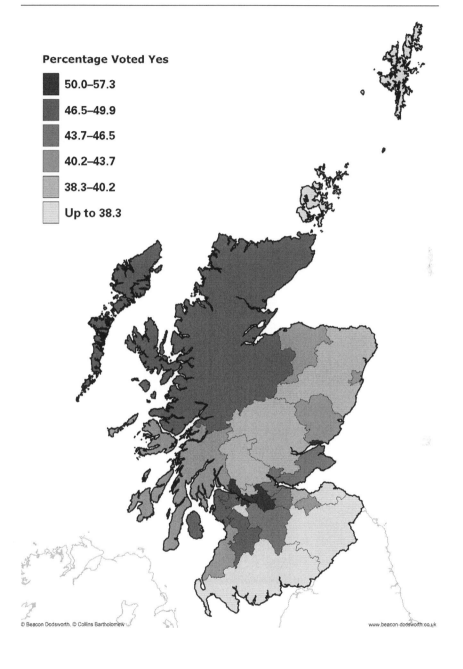

Percentage Voted Yes

- 50.0–57.3
- 46.5–49.9
- 43.7–46.5
- 40.2–43.7
- 38.3–40.2
- Up to 38.3

© Beacon Dodsworth, © Collins Bartholomew   www.beacon-dodsworth.co.uk

TABLE 5.2   Census results for the UK by approximate social grade*

Social Grade	Description	% of UK HRPs aged 16–64
AB	Higher and intermediate managerial, administrative, professional occupations	22.17
C1	Supervisory, clerical and junior managerial, administrative, professional occupations	30.84
C2	Skilled manual occupations	20.94
DE	Semi-skilled and unskilled manual occupations, unemployed and lowest grade occupations	26.05

* For a summary table explaining the different types of social grade:
http://www.ukgeographics.co.uk/blog/social-grade-a-b-c1-c2-d-e
SOURCE: UK Geographics Ltd, Blog at http://ukgeographics.co.uk

The HRP describes the person within a household who has been chosen to describe the characteristics for that household usually based upon their economic characteristics.[3] This classification is heavily used for market research. It is seen as the industry standard. The most common description given to this classification is 'ABC1' as this is the groups of households at whom these industries commonly want to target their products.

Figure 5.2 shows the incidence of people within each LA who are aged 16 to 24 years old and social grade DE. The results here are expressed as index values that describe how likely on average people are to be aged 16 to 24 years old and social grade DE: a value of 100 is what you would expect on average; values of 150 are 1.5 times what you would expect on average while 50 is half what you would expect on average. In reality, values between 90 and 110 could describe 'normal' as well – it is values outside this range that are of greatest interest.

The pattern shown in Figure 5.2 has similarities to Figure 5.1. Again, Dundee City and the areas surrounding Glasgow have high index values. The east/west split is not so marked.

When comparing the top 16 LAs in terms of a 'Yes' vote and having a high proportion of 16–24-year-olds who are social grade DE there is a striking similarity – 14 out of the top 16 LAs match (Table 5.3).

FIGURE 5.2   Incidence of social grade DE and age group 16–24 in Scotland

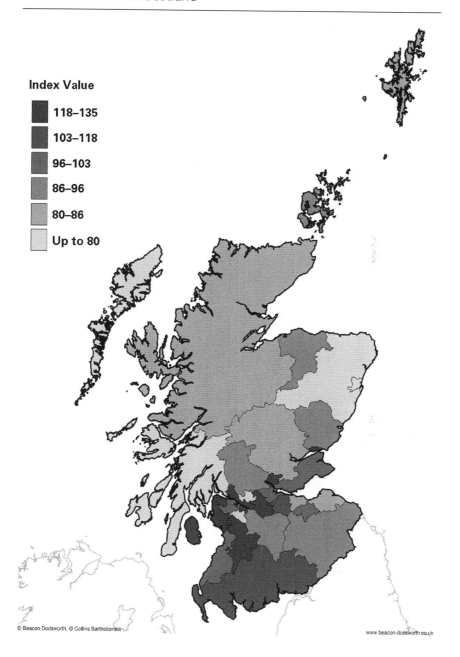

Index Value

- 118–135
- 103–118
- 96–103
- 86–96
- 80–86
- Up to 80

© Beacon Dodsworth, © Collins Bartholomew

www.beacon-dodsworth.co.uk

TABLE 5.3   Rankings of top local authorities on % 'Yes' vote and indexed incidence of 16–24-year-olds in social grade DE

Local Authority	Percent (%) who Voted 'Yes'	16–24-Year-Olds who are Social Grade DE	Index Value
Dundee City	57.35	Glasgow City	135
West Dunbartonshire	53.96	North Ayrshire	128
Glasgow City	53.43	West Dunbartonshire	124
North Lanarkshire	51.07	Dundee City	121
Inverclyde	49.92	East Ayrshire	120
North Ayrshire	48.99	Inverclyde	120
East Ayrshire	47.22	North Lanarkshire	119
Renfrewshire	47.19	Clackmannanshire	119
Highland	47.08	West Lothian	111
Comhairle nan Eilean Siar	46.58	Dumfries & Galloway	108
Falkirk	46.53	Renfrewshire	103
Clackmannanshire	46.20	Midlothian	101
South Lanarkshire	45.33	South Ayrshire	101
Fife	44.95	Falkirk	100
West Lothian	44.82	Fife	97
Midlothian	43.70	South Lanarkshire	96

SOURCE: for voting percentages: http://www.bbc.co.uk/news/events/scotland-decides/results; for index values: Beacon Dodsworth

As is shown in Table 5.3, areas with high levels of 16–24-year-olds of social grade DE do seem more likely to vote 'Yes' to independence. Perhaps it is the young disadvantaged who were most enthusiastic about change as they thought that it would be the best way to improve their lives.[4]

Linking to index values to find a suitable cut-off point, a value of 119 would seem sensible. This would take in the top eight LAs. With only four out of the 32 LAs having a majority 'Yes' vote and a 10 per cent difference in the overall result there is still a little way to go before a change is likely to take place. By the time of the next generation of Scots a 'Yes' vote could happen.

The overall voting turnout was 84.6 per cent. This is extremely high compared to UK general elections over the past 50 years, where the turnout has been 60–65 per cent.[5] Figure 5.3 shows the voting turnout for each LA in the Scottish referendum.

As is shown in Figure 5.3, there is quite a marked variation in turnout percentage across the country. One pattern that does emerge (if compared to Figure 5.1) is that local authorities that voted 'No' to independence tend to have a higher voting turnout. Glasgow had a turnout of 75 per cent compared to Stirling with 91.1 per cent. Would a 10 per cent increase in voter turnout (the overall difference between the 'Yes' and 'No' votes) in Glasgow change the result? This pattern is not perfect, with West Dunbartonshire being an exception to this rule as it had one of the highest voting turnouts: 87.9 per cent. This does suggest that although social grade is an important indicator, it is certainly not the only one. Voter turnout could also contribute.

What about all the talk of extra powers for England? On the back of the Scottish referendum this has started to be discussed by politicians and within the media. To look at this, the same analysis has been carried out to see what the expected patterns might be if there were an equivalent English referendum. For this analysis, it is assumed that there is a very high level of voter interest and turnout would be at similar levels to the Scottish referendum.

Taking an index value of 119 as the cutoff as to whether there would be a majority voting 'Yes', Figure 5.4 suggests that the result is likely to be similar to the Scottish result with the majority of the population not wanting independence. Out of 326 local authorities, 66 (or 28 per cent of the population) fall within this category. It also shows a predominantly north/south split with the traditional heavy industry areas of the Midlands, the north and north-east in favour. These results could back up the idea that areas outside London and the south-east of England would want more power in order to try to address the imbalance of the 'north–south divide'.

FIGURE 5.3   Scottish referendum – percentage turnout by local
authorities

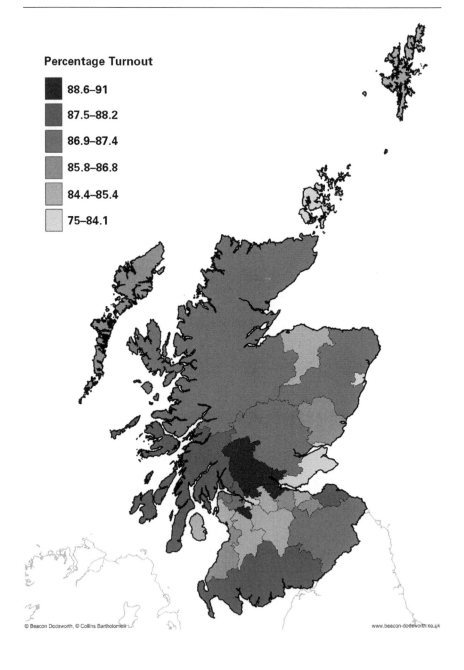

FIGURE 5.4   Incidence of social grade DE and age group 16–24 in England

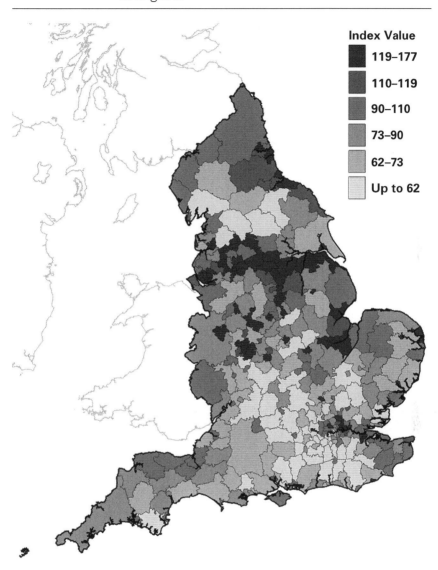

Index Value

- 119–177
- 110–119
- 90–110
- 73–90
- 62–73
- Up to 62

© Beacon Dodsworth, © Collins Bartholomew   www.beacon-dodsworth.co.uk

This example has shown how useful social grade can be when looking into and trying to understand the population of Britain. It is a powerful measure. By combining this with age and sex, many kinds of social issues or studies of a target market can be addressed. Of course when looking into an issue, one variable or indicator is normally not enough to describe a pattern solely on its own – the world is not as simple as that. However, the example gives an insight into the methods that can be used to start to understand a pattern and help answer questions about why a particular outcome has occurred.

Derived indicators

Moving on from analysis of individual variables, census data and other inputs may be used to create indicators that score areas across sets of multiple criteria. Two examples of derived indicators are the Indices of Multiple Deprivation and principal components or factors.

Indices of Multiple Deprivation

Indices of Multiple Deprivation (IMDs) are used in the public sector to rank areas on their levels of deprivation, in order to help target policies and funding. These indices are constructed at the small-area level, based on census data and other sources.

Due to differences between the countries of the UK, four separate indices are created for England, Wales, Scotland and Northern Ireland, following broadly similar methodologies. In each case, levels of deprivation are measured for a number of separate themes, including income, employment, education and health, and are then combined together mathematically using an agreed set of weights. The choice of themes, input statistics within each theme and weights are agreed through consultation with users.

The IMDs are designed to identify the most deprived areas within each country. Therefore, although their values will span the entire range of possible scores, they will be of less use for targeting 'high affluence' areas, which is the more usual goal in product marketing.

There are differences between countries in the definitions of 'small area' employed for the IMDs. The main points of difference are that

England and Wales use lower super-output areas, which have average population sizes of around 1,700 and 1,600 respectively. Scotland uses data zones, on average with 800 people per area, while Northern Ireland uses super-output areas with an average population size of 2,100. For this reason, the Scottish IMD is able to identify smaller pockets of deprivation than the other indices.

Principal components or factors

Another way to summarize a large number of variables is to use multivariate analysis to extract principal components or factors, using PCA (described in Chapter 4). This gives a new set of variables, at neighbourhood level, which express different themes in the original variables.

These factors are an excellent way of scoring each neighbourhood on a set of demographic dimensions. Each factor picks out a particular dimension, such as status, life stage or experience, and so may be used as a discriminator in its own right. The factor scores are typically expressed as either standardized variables or as percentiles.

> **KEY POINT**
>
> These factors can be particularly useful for analysis and modelling, for example as inputs to predictive models, as they take the form of continuous scores.

Example product

A good example of geodemographic factors is the set of UK Mosaic factors from Experian – these distil over 200 variables into five factors, which are available at postcode level. The factor dimensions are as follows:

- Factor A: status.
- Factor B: life stage.
- Factor C: experience.
- Factor D: culture.
- Factor E: work.

Postcode measures of wealth and consumer activity

The Research & Analysis Consultancy (TRAC) has developed a set of postcode scores that measure affluence and consumer activity for all areas in Britain. The scores were derived from a number of sources, including the UK Census, council tax bands, Land Registry house price information and unemployment data. Further information may be found at: **http://www.tracconsultancy.co.uk/index.php?id=wealth-consumer-activity**.

Fuzzy classifications

Neighbourhood classifications tend to be 'crisp' in the sense that each small area belongs either to one segment or another – it cannot belong partly to several segments. Sometimes, this approach might be too clear-cut – a neighbourhood might be close to segment A and also have some characteristics of segment B. Therefore, forcing this locality to belong to segment A could lose this information.

A fuzzy classification retains more of the information, by assigning probabilities of membership to the segments. For example, a neighbourhood might have 95 per cent probability of belonging to segment A and 5 per cent for segment B – in which case a 'crisp' assignment to 'A' would be an accurate description. On the other hand, if it has 40 per cent probability of belonging to segment A and 60 per cent for segment B, then a 'crisp' assignment to 'B' would lose knowledge about the area having some 'A-type' characteristics.

As an example of fuzzy classification, the Luton experiment (overviewed at the end of Chapter 1) was actually analysed by Robin Flowerdew and Zhiqiang Feng using fuzzy clustering. This allowed the three neighbourhoods that did not fit with an expected Acorn type to be examined by their cluster membership probabilities (see Flowerdew and Feng, 1999).

Although the idea of fuzziness in a classification might sound appealing, it introduces an additional layer of complexity in applying geodemographics and so has not been widely adopted in the UK market.

Sub-postcode discriminators

Some of the suppliers of area classifications also 'drill down' below the postcode, and provide segmentations for households and for adult individuals.

KEY POINT

The rationale of sub-postcode discriminators is that an area classifier will, by definition, assign all households and individuals living in any given postcode to the same segment. However, the demographics and circumstances of next-door neighbours are liable to differ and therefore a household-level system can potentially measure these differences, which may be important when targeting the best prospects for a consumer product. Similarly, the adults within a household may have different demographics and needs, which an individual-level classification will seek to measure.

Some of these sub-postcode discriminators are 'general-purpose' products that may be employed in a wide range of markets, while others have been designed with a specific market in mind. As customer profiling and prospect targeting are used widely in financial services, this is a popular sector for market-specific segmentation systems. Table 5.4 provides examples of general-purpose and market-specific systems.

Conclusion

In this chapter we have observed the diversity of discriminators and segmentation systems that exist outside the usual geodemographics heartland of neighbourhood classifications. Information products have been devised to meet all known business applications – we shall start to see why so many different solutions are needed when we review the various business applications in Chapter 7.

TABLE 5.4   Examples of sub-postcode classifications

Supplier	Product	Level*	Target Market
CACI	Household Acorn	H	General purpose
	People UK	I	General purpose
	Fresco	I	Financial services
Callcredit Information Group	CAMEO UK	H	General purpose
	CAMEO Lifestyle	H	Lifestyle segmentation
	CAMEO Property Plus	H	House prices
	CAMEO Choices	H	Lifestyles and purchasing behaviour
	CAMEO UK	I	General purpose
	CAMEO Personal Finance (5 classifications)	I	Financial services
	Connected Technology Segments	I	Online and technology
	Green and Ethical	I	Attitudes to green and ethical issues
Experian	Household Mosaic	H	General purpose
	Individual Mosaic	I	General purpose
	Financial Strategy Segments	I	Financial services

* H = Household, I = Individual

In summary, we have seen that:

- Mapping the patterns in a census variable, such as approximate social grade, can help to understand a more complex outcome, such as the result of the Scottish referendum.

- Areas may be assigned scores that measure their affluence, consumer activity or deprivation levels.

- Fuzzy classification is a way to identify areas that do not fall neatly into a clear-cut neighbourhood type; however, this is rarely used in the UK.

- Sub-postcode discriminators can recognize differences between next-door neighbours, where individual-level targeting is required.

Notes

1 For a summary of what the Census and Geodemographics Group does: https://www.mrs.org.uk/mrs/census_and_geodemographics_group.

2 For the Market Research Society's explanation of social grade: https://www.mrs.org.uk/cgg/social_grade.

3 For a full definition of the household reference person: http://www.scotlandscensus.gov.uk/variables-classification/household-reference-person.

4 For an analysis of the referendum results by Professor John Curtice: http://blog.whatscotlandthinks.org/2014/09/voted-yes-voted/.

5 For the UK general election turnout over the past 50 years, see: http://www.ukpolitical.info/Turnout45.htm.

THE MECHANICS OF USING GEODEMOGRAPHICS

Introduction

This chapter considers the ways in which geodemographic information is usually presented, beginning with the reports that help a user to identify their target segments and then looking at interpretation of the results.

The aims of this chapter are to explain the mechanics of using geodemographics, including:

- Tools and techniques – the various types of geodemographic reports relating to an area, a customer base and a market.
- The types of delivery systems available for users to access and harness geodemographics within their businesses.

- Data visualization methods of greatest relevance for interpreting the results.
- Potential pitfalls and how to avoid them.

For any user, the way the toolkit gets used will depend upon the nature of their business, the particular problems being addressed and the output required in tackling those problems. In Chapter 7 we review the many different applications of geodemographics in a variety of sectors, but first we focus in this chapter on the methods underpinning most of these uses.

Tools and techniques

A number of standard analyses are available from all of the commercial suppliers of geodemographic systems, using their products. To illustrate the typical report formats and explain how to interpret them, a series of examples will be presented using extracts of reports from CACI (equivalent reports are also available from other suppliers).

Area profiles

An area profile summarizes the population characteristics for a specific geographical area, and describes the total potential consumer market available in that region. For companies that operate on a national basis, the area will typically be the entire country; many products, services or brands have a trading area that is, however, geographically restricted in some way – in which case, this area would be profiled. The area may be defined by a variety of methods, including:

- One or more known geographical areas such as regions, counties, local authority districts, postal areas and districts.
- A store catchment area, defined in terms of distance (eg up to five miles from the store) or drive time (eg up to 15 minutes).
- A catchment area created from analysis of the distribution of customer locations around a store.
- A town catchment area that accounts for the majority of trade on either food or non-food products.
- A media consumption area such as local newspaper readership, radio listening or TV region.

The examples shown below focus on a home-furniture store located in Sheffield. A catchment area has been defined, consisting of the postal sectors that account for 95 per cent of non-food shopping in Sheffield. This area is shown in Figure 6.1, and consists of four catchments:

- Primary: the postal sectors that have the highest proportions of spending in Sheffield and jointly account for 50 per cent of the expenditure for the centre.
- Secondary: the postal sectors that contribute the next 25 per cent of expenditure.
- Tertiary: the postal sectors that contribute the next 15 per cent of expenditure.
- Quaternary – the postal sectors that contribute the next 5 per cent of expenditure.

The example area reports in Table 6.1 (census demographics report) and Table 6.2 (geodemographics area report) have been weighted according to the proportion of retail spend from each postal sector that is expected to be made in the Sheffield centre. In effect, the reports, discussed below, therefore estimate the market sizes in terms of numbers of people and households shopping in Sheffield.

Census demographics report

As its name suggests, this report presents a set of demographic profiles of the specified area, and also compares the area against corresponding results for a selected base area. The report helps the user to spot any differences from the given norm and so start to identify any strengths or weaknesses of the area.

Table 6.1 presents an extract from an example report on the Sheffield catchment area for three demographics – gender, age and economic activity – in comparison with a base area of England and Wales. For a business trading in all parts of the UK, the UK would normally be used as the base area. However, at the time of producing this example, comparable detailed 2011 census results were not yet available for the whole of the UK. For this reason, the example took England and Wales as its base.

FIGURE 6.1 Example retail catchment area for Sheffield

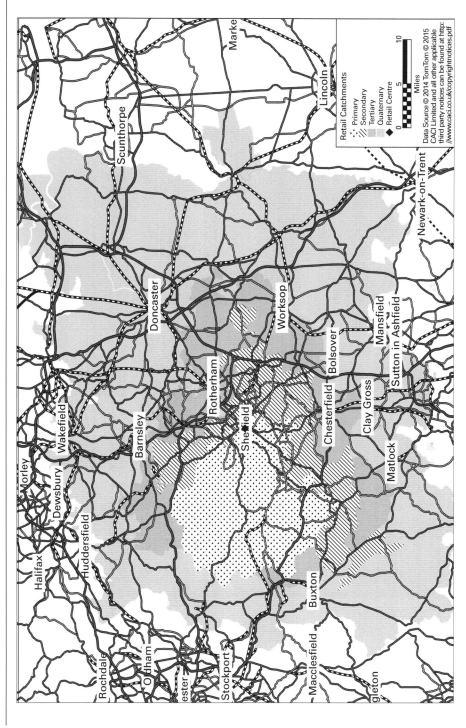

SOURCE: contains Ordnance Survey data © Crown copyright and database right 2014; Royal Mail data © Crown copyright and database right 2014; National Statistics data © Crown copyright and database right 2014

TABLE 6.1   Example 2011 census demographics report

	Profile	Data as % for Area	Data as % for Base	Index
Area: Sheffield Catchment Area				
Base: England & Wales				
Year: 2011				
Total Population	240,694			
Males	118,688	49.3%	49.2%	100
Females	122,006	50.7%	50.8%	100
Age				
0–15	41,871	17.4%	18.9%	92
16–19	15,252	6.3%	5.1%	**125**
20–24	22,833	9.5%	6.8%	**140**
25–44	62,055	25.8%	27.4%	94
45–64	58,594	24.3%	25.4%	96
65 +	40,087	16.7%	16.4%	**101**
All People aged 16 to 74	**179,936**			
Economically active	119,061	66.2%	69.7%	95
Employee	87,636	48.7%	52.2%	93
Self-employed	14,456	8.0%	9.7%	83
Unemployed	7,721	4.3%	4.4%	98
Full-time student economically active	9,248	5.1%	3.4%	**150**
Economically inactive	60,874	33.8%	30.3%	**112**
Retired	25,551	14.2%	13.8%	**103**
Other economically inactive	35,323	19.6%	16.5%	**119**

SOURCE: © Crown Copyright 2012; Office for National Statistics, Licensed under the Open Government Licence v.1.o; © CACI Ltd 2013

TABLE 6.2 Example area profile report by Acorn groups

Area: Sheffield
Base: United Kingdom
Year: 2014

Acorn Group Description	Profile	Area %	Base %	Index	0	100	200
1. Affluent Achievers							
1A Lavish Lifestyles	1,353	1.3	1.2	114			
1B Executive Wealth	12,168	11.8	11.4	104			
1C Mature Money	11,728	11.4	9.4	121			
2. Rising Prosperity							
2D City Sophisticates	606	0.6	3.4	17			
2E Career Climbers	2,670	2.6	5.9	44			
3. Comfortable Communities							
3F Countryside Communities	3,656	3.6	6.4	55			
3G Successful Suburbs	4,876	4.7	5.8	82			
3H Steady Neighbourhoods	8,092	7.9	7.8	101			
3I Comfortable Seniors	2,745	2.7	3.1	85			
3J Starting Out	6,153	6.0	4.0	150			
4. Financially Stretched							
4K Student Life	6,837	6.6	2.0	339			
4L Modest Means	5,985	5.8	7.5	78			
4M Striving Families	6,675	6.5	7.8	83			
4N Poorer Pensioners	9,712	9.4	5.8	162			
5. Urban Adversity							
5O Young Hardship	3,701	3.6	5.7	63			
5P Struggling Estates	6,042	5.9	7.1	83			
5Q Difficult Circumstances	9,716	9.4	5.5	173			
6. Not Private Households							
6R Not Private Households	205	0.2	0.2	83			
Total	102,920						

The census demographics report provides:

- the area market size, in total and by each profile group;
- the proportion of the area's market having each characteristic;
- the corresponding proportion of the base area having each characteristic;
- the index value for each characteristic.

The index shows how the proportion of each demographic attribute for the area compares with the equivalent figure in the base. Differences between these two values are measured by the index in the following way:

- An index of 100 indicates that the proportion within the area is the same as that in the base.
- An index of over 100 shows an above-average representation (eg 140 shows that the area proportion is 40 per cent greater than the equivalent item in the base).
- An index of under 100 shows a below-average representation (eg 60 shows that the area proportion is 40 per cent below the equivalent item in the base).

The example report shows that the area has a total market size of just over 240,000, of whom 180,000 are aged 16–74. Age groups 16–19 and 20–24 are over-represented (in comparison with England and Wales) and together account for nearly 16 per cent of the area. The area has 50 per cent more full-time economically active students than the proportion in the base area, and also 19 per cent more economically inactive people who are not retired.

   A full census demographics report provides a lot more information than this small example has shown, including additional profiles on attributes such as family composition, housing tenure, car ownership, travel to work, occupation, National Statistics socio-economic classification (SEC), ethnicity and religion.

Geodemographic area report

The geodemographic area report presents the composition of an area by geodemographic types, compares these results against equivalent figures for a chosen base area and presents the comparison as an index, as defined above.

As an illustration, Table 6.2 provides an Acorn group profile of the Sheffield area, against a UK base. The 'Profile' column presents the household count for each Acorn group, totalling nearly 103,000 households in the entire catchment area. The 'Area %' column shows these household counts as percentages, for comparison against the corresponding percentages for the 'Base'. The 'Index' column directly compares the 'Area' and 'Base' percentages, in the same way as for the census demographic report . These index values are displayed graphically as a series of bars, in order to help the reader to identify Acorn groups that are particularly 'strong' or 'weak' in that area.

So, for example, we see that Group 3J (Starting Out), while only 6 per cent of the area, is over-represented in the Sheffield catchment with an index value of 150. Similarly, groups 4K (Student Life), 4N (Poorer Pensioners) and 5Q (Difficult Circumstances) are also strongly over-represented. A more valuable group will probably be 1C (Mature Money), accounting for 11 per cent of households – this group is over-represented by 21 per cent (according to its index of 121). In contrast, the area contains relatively few households in the groups within Rising Prosperity (groups 2D and 2E in total have 3.2 per cent in the area, versus 9.3 per cent in the base).

A full geodemographic report provides all levels of the classification – for this example starting with the high-level Acorn category profile, then drilling down into the Acorn groups analysis as shown, and finishing with Acorn types at the granular level.

Customer profiles

By appending geodemographic codes on to a file of postcoded customer records, a geodemographic customer profile may be produced – either for the entire customer base, or for a subgroup such as those shopping in a particular store, buyers of certain products, high-value customers, and so on.

The geodemographic customer profile compares a group of customers against a suitable base, typically the profile of all households in the catchment area.

The Acorn group example presented in Table 6.3 profiles a sample of customers for the Sheffield furniture store, against a base of all households in the Sheffield catchment area. In this case, the 'Base' has not been weighted to represent consumer spending in Sheffield, but shows the total numbers of households in that region. The report includes the following columns:

- The 'Profile' count and percentage (number and proportion of customers in each Acorn group).

- The corresponding household count and percentage for the 'Base' area and the 'Penetration' percentage, ie the proportion of customers within each Acorn group.

- The 'Z-score', which measures the statistical confidence that may be placed in the difference between the 'Profile' and 'Base' percentages. As a rule of thumb, if the Z-score is either greater than +2 or less than –2, then the difference is viewed as being significant and so is likely to be real (rather than due to sampling variation).

- The 'Index' value for the 'Profile' per cent compared with the 'Base' per cent, as defined above, together with a corresponding bar chart to identify high and low indices.

If we compare the bar chart in Table 6.3 with the previous example in Table 6.2, we immediately see a very different pattern of results. These show that the Sheffield customers belong more to the 'Affluent Achievers' and 'Rising Prosperity' categories and their constituent Acorn groups. Their Z-scores are all greater than 2, so these differences are likely to be real. Other groups of importance include 3J (Starting Out), 3H (Steady Neighbourhoods) and 5Q (Difficult Circumstances).

Again, a full customer profile report will include all levels of the classification, enabling interesting findings at the 'groups' level, such as the high index on group 5Q, to be examined for the types within that group.

KEY POINT

Having profiled all customers and identified their geodemographic strengths and weaknesses, further insights may often be gained by carrying out profiling for subsets of interest in comparison with 'all customers' as the base. For example, this could be used to look for differences in characteristics between repeat versus one-time buyers or buyers of different products.

TABLE 6.3 Example customer profile report by Acorn groups

Profile Title: Sheffield Customers
Base Title: Sheffield Retail Footprint

Acorn Group Description	Profile	%	Base	%	Penetration %	Z-Score	Index	0 · · · 100 · · · 200
1. Affluent Achievers								
1A Lavish Lifestyles	39	1.1	3,993	0.4	1.0	7	298	
1B Executive Wealth	402	11.5	97,208	9.1	0.4	5	126	
1C Mature Money	369	10.5	94,470	8.8	0.4	4	119	
2. Rising Prosperity								
2D City Sophisticates	14	0.4	1,596	0.1	0.9	4	268	
2E Career Climbers	111	3.2	24,134	2.3	0.5	4	140	
3. Comfortable Communities								
3F Countryside Communities	53	1.5	77,793	7.3	0.1	-13	21	
3G Successful Suburbs	171	4.9	69,640	6.5	0.2	-4	75	
3H Steady Neighbourhoods	376	10.7	88,932	8.3	0.4	5	129	
3I Comfortable Seniors	79	2.3	37,501	3.5	0.2	-4	64	
3J Starting Out	316	9.0	39,890	3.7	0.8	17	242	
4. Financially Stretched								
4K Student Life	99	2.8	18,229	1.7	0.5	5	166	
4L Modest Means	215	6.1	100,999	9.4	0.2	-7	65	
4M Striving Families	249	7.1	92,482	8.6	0.3	-3	82	
4N Poorer Pensioners	243	6.9	133,505	12.5	0.2	-10	56	
5. Urban Adversity								
5O Young Hardship	151	4.3	54,058	5.1	0.3	-2	85	
5P Struggling Estates	239	6.8	73,647	6.9	0.3	0	99	
5Q Difficult Circumstances	371	10.6	59,564	5.6	0.6	13	190	
6. Not Private Households								
6R Not Private Households	8	0.2	2,171	0.2	0.4	0	112	
Total	3,505		1,069,812		0.3			

Market research profiles

There has always been a strong linkage between geodemographics and market research – this goes back to the MRS Conference in 1979, where it was demonstrated that a neighbourhood classification overlaid on to the Target Group Index (TGI) produced exciting new insights.

The larger syndicated research surveys are coded by geodemographic systems, enabling users to obtain reports and analyses for their desired markets – either at total market level, or by sectors and brands. These profiles can be invaluable for products where customer details are not recorded or would be costly to obtain, such as recent entrants to a market. Research profiles are also a useful source of market involvement rates (or penetrations) by neighbourhood types for estimating market sizes in small areas (see 'Small-area estimation' below). Table 6.4 lists some of the industry surveys from which geodemographic profiles are available – these cover purchasing of consumer products, media usage and other lifestyles and behaviours.

Continuing the previous example, Table 6.5 shows an example TGI profile for purchasing of household furniture, fittings and furnishings, analysed by Acorn groups. The format of the report is very similar to the customer profile (Table 6.3); however, the 'Profile' and '%' columns now represent the buyers of those products within the TGI survey, while the 'Base' and '%' columns represent the total TGI sample. The 'Penetration %' column compares the 'Profile' with the 'Base' and so represents the incidence of purchasers within each Acorn group. The 'Index' column compares these penetrations with the overall penetration for household fittings and furnishings (21.1 per cent) to identify which groups are relatively more (or less) likely to contain purchasers – the index values are also displayed as bar charts for ease of interpretation. Finally, the Z-score (explained earlier) provides a measure of the statistical significance of each index.

There are some similarities and some differences between the TGI profile shown in Table 6.5 and the customer profile shown in Table 6.3. In both reports, groups 2D ('City Sophisticates') and 2E ('Career Climbers') purchase at above-average rates. Similarly groups 1B ('Executive Wealth'), 3J ('Starting Out') and 4K ('Student Life') are important in both. However, the other groups that were 'strong' in the customer profile do not feature markedly in the TGI report. Overall, if we were to plot the two sets of

TABLE 6.4 Some sources of market research profiles

Topic	Supplier	Survey	Website
Product and media usage	Kantar Media	Target Group Index (TGI)	www.kantarmedia.com
Readership and media usage	Ipsos MORI	National Readership Survey*	www.nrs.co.uk
FMCG products	Kantar	Worldpanel	www.kantarworldpanel.com
FMCG products	Nielsen	Homescan panel	www.nielsen.com
Financial services	GfK	Financial Research Survey (FRS)	www.gfk.com
Financial services	Ipsos MORI	Ipsos MORI Financial Services	www.ipsos-mori.com
Demographics and lifestyles	DataTalk Research	British Population Survey	www.thebps.co.uk

* A survey of newspaper and magazine readership in Britain

TABLE 6.5   Example TGI profile report by Acorn groups

Profile Title: Household Furniture, Fittings and Furnishings
Base Title: Total Sample

Acorn Group Description	Profile	%	Base	%	Penetration %	Z-Score	Index	0 100 200
1. Affluent Achievers								
1A Lavish Lifestyles	21	0.3	102	0.4	20.6	0	98	
1B Executive Wealth	639	10.5	2,185	7.6	29.2	9	138	
1C Mature Money	523	8.6	2,332	8.1	22.4	1	106	
2. Rising Prosperity								
2D City Sophisticates	193	3.2	587	2.0	32.8	6	155	
2E Career Climbers	509	8.4	1,601	5.5	31.8	10	151	
3. Comfortable Communities								
3F Countryside Communities	271	4.4	1,464	5.1	18.5	-2	88	
3G Successful Suburbs	433	7.1	1,790	6.2	24.2	3	114	
3H Steady Neighbourhoods	497	8.1	2,346	8.1	21.2	0	100	
3I Comfortable Seniors	134	2.2	837	2.9	16.0	-3	76	
3J Starting Out	452	7.4	1,574	5.5	28.7	7	136	
4. Financially Stretched								
4K Student Life	144	2.4	553	1.9	26.0	3	123	
4L Modest Means	511	8.4	2,769	9.6	18.5	-3	87	
4M Striving Families	433	7.1	2,742	9.5	15.8	-6	75	
4N Poorer Pensioners	247	4.1	1,824	6.3	13.6	-7	64	
5. Urban Adversity								
5O Young Hardship	388	6.4	2,047	7.1	19.0	-2	90	
5P Struggling Estates	407	6.7	2,188	7.6	18.6	-3	88	
5Q Difficult Circumstances	290	4.7	1,882	6.5	15.4	-6	73	
6. Not Private Households								
6R Not Private Households	7	0.1	40	0.1	17.9	0	85	
Total	6,099		28,863		21.1			

indices against one another, we would see a strong linear relationship between them, with a correlation of 52 per cent.

Small-area estimation

The 'Penetration %' column in Table 6.5 shows the incidence of buyers within each Acorn group. By applying these penetrations to population sizes by Acorn groups, for an area, we can estimate its total market size. This forms the basis for a straightforward method of estimating market sizes at small-area level, employed by many of the geodemographic suppliers (small-area estimation is discussed further in Chapter 10).

Accessing geodemographic information

The routes that users follow to obtain geodemographic information depend upon their business requirements and their analytical capabilities. This section identifies three main options – directory products, analytical systems and bureau services.

Directory products

Some users will simply want to append a geodemographic classification code on to each of their customer or prospect records, and use it as a data attribute within their in-house systems, eg for campaign analysis and response modelling. This can be achieved by obtaining a postcode look-up file, or directory, giving the classification value for each unit postcode in the UK. The various suppliers listed in Chapter 4 distribute postcode directories and regular updates to their clients, as a standard process.

Analytical systems

Analytical systems enable users to carry out tasks such as geographical analysis, geodemographic profiling and mapping for themselves. There are broadly two levels of solution available here – geographic information systems (GIS) and geographical analysis systems – the distinction between them is discussed below. Table 6.6 provides examples of some of the products currently available under each of these headings.

TABLE 6.6 Examples of GIS products and geographical analysis systems

GIS Products	Supplier	Website
ArcGIS	ESRI	http://www.esri.com/
Infrastructure Map Server	Autodesk	http://www.autodesk.com/
GeoMedia Professional	Intergraph	http://www.intergraph.com/
MapInfo Professional	MapInfo	http://www.mapinfo.com/
Clark Labs	IDRISI	http://www.clarklabs.org/
Caliper	Maptitude	http://www.caliper.com/
GRASS (open-source GIS)	GRASS GIS project	http://grass.osgeo.org/
Geographical Analysis Systems	**Supplier**	**Website**
Micromarketer, Location Analyst	Experian	http://www.experian.co.uk/
InSite	CACI	http://www.caci.co.uk/
MICROVISION	Callcredit	http://www.callcredit.co.uk/

Geographic information systems (GIS)

GIS software is primarily used to create, manage, analyse and map geographical data. Typical applications for GIS include store location analysis, vehicle routing and network management by utilities. Different categories of GIS software exist to deliver the various functions required by the wide range of GIS applications – the main distinction lies between desktop GIS products and online products. There is a substantial body of literature on GIS – see, for example, Burrough and McDonnell (1998), DeMers (2008), Heywood, Cornelius and Carver (2011), and Longley *et al* (2005).

Geographical analysis systems

The full functionality of GIS software may not be needed for producing the kinds of geodemographic analyses that were illustrated earlier in this chapter. Geographical analysis systems are geared to importing data and producing geodemographic or demographic reports and maps. The most straightforward of these tools are available online, while desktop solutions also exist.

Bureau services

The geodemographic suppliers provide bureau services to satisfy more ad hoc or bespoke requirements. This route is also more appropriate for application of individual-level discriminators, since customer name and address data will often have to be standardized to the supplier's own format.

Visual presentation of results

Nowadays we are all familiar with on-screen maps, displaying the results of our internet searches or showing us recommended routes for driving from 'A' to 'B'. Therefore, maps are an obvious medium for presenting geodemographic results visually.

Similarly, we expect to see profile results presented in a visual format, where possible, as an aid to interpretation. Geodemographics suppliers have continuously improved their online and desktop systems to present information in a more accessible format. For example, the supporting materials usually available on neighbourhood classifications combine a wide range of ready-made profiles with excellent visualization tools.

The Personicx Eye interactive tool from Acxiom (see **http://www.personicx.co.uk/**) enables the user to visualize the target types for a wide range of different target audiences and to gain greater insights about those segments.

Another example is the Segmentation Portal from Experian, which is designed as an interactive guide for the Mosaic UK classification. This includes several innovative graphical displays, such as the family tree and the custom family tree.

The family tree, illustrated in Figure 6.2 for the current Mosaic UK classification, displays the relationships between the segments and is another example of the minimum spanning tree introduced in Chapter 4. Segments that are most similar to one another are joined together by lines in the tree graph – to suggest either which other segments are close to a given target segment, or how people are likely to migrate between segments.

FIGURE 6.2   Example family tree for Mosaic UK

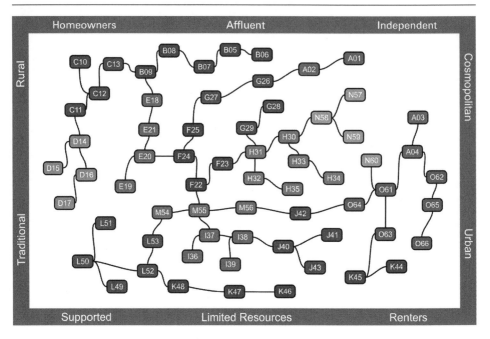

SOURCE: Reproduced with the permission of Experian

The custom family tree plots the positions of Mosaic types or groups, based on a pair of variables selected by the user, in order to identify target segments based on multiple criteria. The example shown in Figure 6.3 is based on household income and smart TV ownership – to show which Mosaic types are likely to be in the market for purchasing a smart TV and how strongly the classification discriminates for this audience. This chart suggests a positive correlation between the two axes: for example, Mosaic types with highest average household incomes, B07, B05 and A01, are more likely to own a smart TV.

KEY POINT

When exploring any neighbourhood classification for the first time, a good way to start is by entering a few known postcodes into the system – in order to find out how these are classified and whether the descriptions make sense.

FIGURE 6.3 Example custom family tree for Mosaic UK

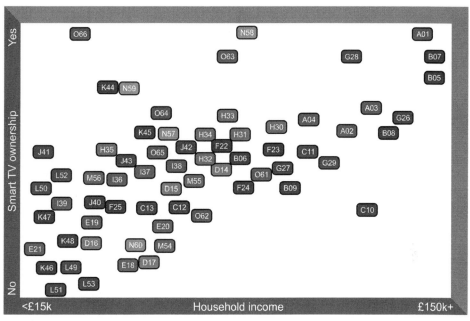

SOURCE: Reproduced with the permission of Experian

Personal example

When starting to explore the Mosaic classification, I entered my own postcode and learnt that I was in type B05 – 'Premium Fortunes'. I was planning to downsize from a house to a flat at that time, and also entered my new location – type B06, Diamond Days. The interesting point was, according to the family tree (Figure 6.2), that types B05 and B06 are directly connected; therefore my planned move was completely logical. However, the family tree shows one slightly disconcerting feature – type B06 lies at the end of the line and suggests that I will have nowhere else to go from there! However, this only means that no other type is sufficiently close to B06, to be connected with it in the tree.

Potential pitfalls and how to avoid them

There are a couple of traps for the unwary when using geodemographics, which can arise when interpreting neighbourhood profiles or analysing area data. These are the ecological fallacy and the modifiable areal unit problem – they are explained in turn below, together with suggestions for how avoid them.

Ecological fallacy

As we saw in Chapter 4, neighbourhood classifiers are derived from area profiles; however, they are often employed to infer characteristics of individuals. The ecological fallacy is the false assumption that these inferences will always be correct. The reason why this assumption is false is that area profiles summarize average values for areas – in one area, all the households may be similar to one another and will be well represented by that average, but in another area the households may vary more and will be poorly represented by the average.

> **KEY POINT**
>
> The ecological fallacy can manifest itself in several ways. The first, and most dangerous for the inexperienced user, is that the segment assigned to a neighbourhood will not necessarily be accurate for every household in that area. So, for example, neighbourhoods classified as wealthy can contain poor people and vice versa. Therefore, it is essential to keep in mind that the geodemographic segment is a label based on an area profile – rather than a precise classification of each individual. By definition, this fallacy should not apply to the individual-level discriminators described in Chapter 5, provided that they are fully based on individual attributes.

Another instance of the ecological fallacy arises if census profiles are produced based on a set of customer postcodes. This type of profile is sometimes used to look at demographic characteristics for a list of postcodes – it operates by assigning each postcode to its census output area, extracting the area's census profile and then reporting the aggregation of those profiles across all of the output areas. Provided that the results are interpreted with caution, in comparison with an appropriate base area, they can help to identify any strong customer characteristics. However, they can also lead to incorrect conclusions. For example, a researcher took several thousand geocoded tweets, linked each to an output area and produced a census profile. The resulting analysis contained a large number of census variables that could easily be misinterpreted – for example, it included children aged 0–4, who were clearly too young to tweet!

Modifiable areal unit problem

The modifiable areal unit problem (MAUP) can arise when individuals are grouped into geographical areas for reporting and analysis. There are two components of MAUP, both arising from overlaying geographical boundaries on to location data – these are effects of scale and of zoning.

Effects of scale

The results of any geographical analysis are likely to be influenced by the scale of the areal units that are employed. For example, if measuring the incidence of a demographic group reported by the population census, different results are likely to be obtained for different units – be they regions, local authority districts or output areas. For this reason, careful thought should be given at the outset, in order to select the most appropriate geography for addressing the business problem on hand. If in doubt, it is better to conduct analysis at a more granular geographical level than might be required – the results can always be aggregated to higher geographies, but can never be disaggregated in the other direction. Also, all the units in the analysis should, if possible, be at the same scale – avoid unnecessarily analysing together areas that are at different geographical scales.

Effects of zoning

The choice of zones or geographic units can affect the aggregated results, even if the areas are all of the same size. The same problem often occurs in politics, where it is known as gerrymandering, ie manipulating an electoral area's shape for political gain. Geodemographics users typically work with a given predefined geography, in which case there is little that can be done about zoning effects; however, users should still consider whether important results are real or possibly due to the nature of the geographical boundaries.

Conclusion

We have seen that census data and geodemographic classifications can be used to describe retail catchments and other geographies, such as media areas, towns and regions. Similarly, geodemographic profiling may be applied to postcodes of customers and prospects in order to help understand their characteristics. Where customer data cannot be used, or if we want to take a market-wide perspective, then profiles can be obtained from a variety of syndicated market research surveys covering different vertical markets.

A variety of delivery methods is available for accessing geodemographic information, including desktop geographic information and analysis systems for client use, and bureau services for more bespoke requirements.

Data visualization techniques are invaluable for helping to interpret geodemographic results – particularly geographical maps for displaying spatial patterns and graphical techniques for summarizing relationships between segments.

Finally, when interpreting profiles based on neighbourhood data, users should keep in mind that a geodemographic description will not necessarily be accurate for everyone who has been assigned that label. And the results of analysing area data may depend upon the geographical units that are being used.

APPLICATIONS IN VARIOUS INDUSTRY SECTORS

07

Introduction

The knowledge and insights from geodemographics have been harnessed in a variety of ways by organizations with customers to manage, citizens to serve or consumers to research. This chapter describes some of these applications.

We start by reviewing general applications and the benefits of deploying geodemographics, as a tool that combines segmentation with geography. After that, a series of industry sectors are considered in turn for some of the more specific ways in which they draw on this information. Contributions and case studies are included from practitioners and suppliers working in different fields in order to explain how their organizations are applying the science.

General applications

Geodemographics can deliver all the applications and benefits of customer segmentation, enhanced by the unique ability to locate the segments 'on the ground'. As marketers know well, segmentation can be a valuable tool for recruiting and managing customers. By segmenting a market, appropriate products and services may be offered to each customer, via relevant channels and in suitable styles and tones of voice.

The preceding chapters have shown that geodemographic segmentation provides additional benefits – a known segment profile for every postcode, household and individual or, in other words, the ability to identify people who belong to any target segment.

Linking datasets together

A general use of geodemographics is for linking datasets together by segments, avoiding any need for large volumes of data to be moved or physically matched. For example, a market research survey can be profiled to identify target segments for a product or service and then those segments may be identified within a customer database. Likewise, the corresponding segment profile can be obtained for readers of different newspapers and compared with the product purchase profile – to decide where to place press ads. Alternatively, the research source can quantify purchase rates by segments, which may be overlaid on to the profile of a store catchment in order to predict demand in that area.

Applications such as these were first devised and implemented in the 1980s by practitioners involved in marketing and advertising in the private sector. Many of the same techniques are still being applied some 30 years later, within automated systems. Corresponding uses also apply in the public sector and in academia, where the adoption of geodemographics has been increasing over recent years.

KEY POINT

Creating a holistic view of each customer

Extending the idea of linking datasets together, geodemographic segments can help to form a holistic view of each customer, in terms of their market-wide purchasing. Provided that a research source is available that measures the total market, then first use this source to calculate the average market-wide expenditure within each segment. Provided that you know the spend on your brand by each of your customers and which segment each belongs to, then you can employ the average market expenditure for that segment as the 'overall wallet size' for the customer. Hence you can derive your share of wallet by customer and examine how this relates to other factors such as satisfaction with your brand. This can also enable you to identify which customers would be your best prospects for increasing their spend or purchasing additional products.

The 'geo' component implies that geodemographic segments are locatable and therefore more actionable than demographic, attitudinal or behavioural segmentations. So, for example, postal sectors may be ranked on the incidence of a target geodemographic type identified from a research profile, so that door-to-door leaflet distribution can be targeted to the best areas. Similarly, market researchers who need to reach minority populations can target the locations that are most likely to contain those groups, and therefore reduce their fieldwork costs.

Of course, there is no perfect segmentation that can correctly identify who will buy a product and who will not. As with any approach, some purchases will come from a mixture of unlikely segments that can never be predicted. The main point of geodemographics, along with other targeting tools, is to increase the chance of each contact being successful. The targeting only needs to make a small improvement over a random selection and it will generate significant gains in profitability or efficiency.

Marketing applications

Given its flexibility and ease of use, there is really no limit to the potential applications of geodemographics in marketing. Some of its uses apply widely, across many different industries and sectors – these are identified and outlined briefly below.

Helping to understand your customers better

Profiling your own database is an ideal way to start learning more about your customers, and to identify which geodemographic segments are most important to your business. Extending the principle, subsets of your customer base may also be profiled, to identify how each group differs from your overall customer profile. For example, depending on your business, it may be useful to look at subsets such as:

- responders to your different types of marketing campaign;
- buyers of the different brands, products and services available from your company;
- customers who purchase via different channels, such as direct mail versus press advertising versus online;
- customers who transact via different channels, such as in-branch versus online;
- your own customer segments, and/or subsets of customers based on value to your business.

You will also want to understand how your customers differ from those of your competitors and from the market as a whole – this implies profiling external sources that cover your market in sufficient detail. The ideal source would be a market research survey that identifies buyers of the total market and each main brand, including your own. All of these buyer groups should be profiled and interpreted in comparison with your internal set of profiles.

Targeting customer campaigns

Geodemographics can be a useful ally if your company sends out targeted campaigns, such as for cross-selling products or for customer retention.

The data that you already hold for each customer, such as their transaction history, will doubtless provide the most powerful selection criteria and may be combined into a targeting model for each type of campaign. Geodemographic segments can sometimes improve the selection or the model, in two possible ways.

First, geodemographics provides a context for each of your customers, painting a picture about the circumstances surrounding them, as Richard Webber discussed in Chapter 2. Second, if your product campaign is likely to be most successful within a specific demographic group, then the corresponding geodemographic segment is likely to be a useful selection criterion or variable in your model.

Targeting external list selections

If you are going outside your customer base to recruit new customers from an external list, then you are likely to find that a much smaller number of selection criteria are available to you. Geodemographic segments are normally available amongst these criteria and so should help you to target the priority segments for your business.

Targeting door-to-door distribution

Door-to-door distribution is commonly targeted to small areas such as postal sectors; it is a more cost-effective way to deliver messages to households in a postal sector than using direct mail. Your target geodemographic segment may be used to rank the available postal sectors, either for a large-scale campaign, or for a local campaign within a defined catchment area. Therefore, the door-drop may be designed to reach the largest number of people in your target segment.

Targeting online advertising

Much of the advertising that we see when visiting a website is targeted in real time by demand-side platforms (DSPs). These DSPs interrogate data-management platforms (DMPs) in order to determine characteristics of site visitors. The DSP then selects each advertisement to display out of all ads available to it, using selection criteria defined by the advertising agencies, and instructs the website to display it. All of this takes place in a fraction of a second, as the web page loads.

Although the selection criteria are often behavioural (eg known interests from recent browsing activity), geodemographic segments may be included in the process. Geodemographic suppliers have formed agreements with DMP owners to include their segmentations on DMP databases. The DSP can include geodemographic segments amongst its selection criteria and can choose whether to select a cookie and display an advert to a recipient based on their segment membership. Each time this occurs, the DMP owner pays a fee to the geodemographic agency.

The remainder of this chapter illustrates some of the more industry-specific ways in which geodemographic techniques are applied in different sectors.

Retail applications

Geodemographics is extensively used by retailers with stores or branches to manage, due to the obvious importance of the local population to the performance of a store. Location analysis can help to answer questions such as:

- What is the catchment area of a retail outlet and the characteristics of the population served within that area?
- Where should new outlets be sited, and how well are existing outlets performing?
- What merchandise range should be stocked in an outlet, in order to meet the needs of its catchment area population?
- Which local media should be used to market a store within its catchment area?

Geodemographics can be a valuable component in addressing each of these issues. Its main retail applications are set out below.

Catchment area analysis

Retailers use several alternative methods for defining catchment areas, dependent mainly on whether the outlet is already trading or a location being considered for a new store. For an existing store, customer postcodes may be captured - either obtained from transaction data or by conducting a short survey amongst a sample of shoppers as they exit the store. A catchment area may then be constructed, either by mapping the postcodes and defining an area that supplies the majority of spending to the store, or by identifying the postal sectors that account for most of the trade (as in the Sheffield example in Chapter 6). For a potential new store, a catchment area may be defined in terms of distance or drive time around the location – perhaps assessed from analysis of existing stores – and then collecting up the postal sectors that fall wholly or partly within this area.

Once the catchment has been defined, the retailer will typically require information about the area, such as:

- The total number of people and households residing in the area.

- The demographics of the resident population – particularly the incidence of any characteristics that are key drivers for that retailer's business.

- The geodemographic profile of the area, particularly the presence of any known target segments.

- The expected total demand for that area on the markets served by that retailer.

Some of this information may come directly from census data, while other elements come from the geodemographic profile of the area. The catchment area example, at the start of Chapter 6, demonstrates catchment area analysis in action.

Site location

The investment required for opening a new store is so great that many large retailers employ site location analysts to inform the decision-making process and reduce the likelihood of 'getting it wrong'. An initial site-screening process is often used to make quick assessments of sites, or to generate locations that satisfy initial selection criteria, such as catchment area population size, quality or potential spend.

A more detailed study will then follow for shortlisted sites, producing revenue and market-share forecasts that take into account presence of competing outlets serving the same catchment area. The largest retailers operate site location models for predicting the future performance of a new site. Some of these use gravity models (also known as spatial interaction models) to analyse and predict store turnover and the impact of opening/closing competing outlets. Others use regression models for estimating branch performance based on the key characteristics of the outlet and its catchment area. Knowledge of the catchment area and its retail potential are essential inputs to all of these methods, and these will be sourced from geodemographics, either directly or indirectly.

For readers interested in store location techniques, these methods are discussed in detail by Davies and Rogers (1984) and Wrigley (1988).

Merchandise ranging

Merchandise ranging may be planned with the help of geodemographics. By combining the segment mix of the catchment area with purchasing rates by segments, for each of the products offered by the retailer, the potential demand may be estimated for each product. The retailer can therefore select its range at store level in line with the predicted demand profile.

Local store marketing

Channels such as local press, radio, posters, direct mail and leaflet distribution may be planned and targeted with the help of geodemographics, for example in order to support store openings and relaunches, or to defend against competitive activity. Applications of geodemographics to these channels are discussed later in this chapter.

The following case study describes how potential town locations were assessed and prioritized for a prospective furniture retailer in the south of England.

CASE STUDY   Retail (provided by Beacon Dodsworth)

Objective

Beacon Dodsworth client Paul Elson was looking to site his first furniture outlet in the south of England. He had access to existing retail data from the Bristol area in order to create a model and analysis of potential new retail locations. The client needed a ranking of the best locations within his study area, together with other data such as estimates of retail potential for the furniture market.

Scenario

The client required a formal report with supporting data to indicate a good starting location for furniture retail premises. He suggested a set of possible targets within a study area ranging from the south coast to the M4 and from Wimborne Minster to Royal Tunbridge Wells. Retail figures from existing stores in the Bristol area were provided.

Analysis

- Starting data

  Average weekly retail figures were supplied for five locations between Weston-super-Mare and Gloucester. Seasonal variation in furniture spend was illustrated by providing figures for March to September and October to February.

  Client experience suggested that most customers will travel up to 20 minutes to reach a store. The client also identified the target P^2 People and Places tree types of the source store customers. Also provided was a publicly available list of competitor sites.

- External data sources

  The potential value of the furniture market was based on the data from the Living Costs and Food (LCF) survey, indicating weekly spend per household on furniture and furnishings by P^2 People and Places tree types.

- Target locations

  The client identified 42 possible towns within the study area. These towns were geocoded to identify the centre of each settlement.

- Customer model

  The client supplied data for weekly takings and catchment areas for the stores in the Bristol area. Combining the information with the LCF survey data produced a total possible weekly spend on furniture and furnishings in the catchment area as well as the estimated market shares of the individual stores.

  The source store catchments were defined using 20-minute drive times. Household counts in the catchment for the P^2 People and Places tree types identified by the customer as their core target were combined with the LCF data to produce a total potential weekly spend for each catchment area.

  Beacon Dodsworth generated a market-share estimate by matching the potential spend to the recorded sales figures from the stores. The worst, average and best cases were produced, including seasonal variation and annual averages.

- Retail potential
  A series of catchments for the 42 towns was generated using 20-minute off-peak driving times around their locations. Using the P^2 profile and LCF data, the total market for each catchment was estimated. The retail potential for each target location was calculated by applying the market penetration percentage to the total market size.

Benefits

The work delivered an ordered list of potential sites. The client used the report and supporting documents in the preparation of business plans. The analysis used publicly available datasets alongside customer-specific data. The combination of specific and general data produced a report that matched the client's expectations where he had local knowledge.

Financial services applications

Companies in the financial services sector use many of the marketing applications of geodemographics – the sector is rich in having both customer data and market research. In addition, those providers operating branch networks apply equivalent techniques to the retail sector, for planning and evaluating their branch locations.

Using geodemographics for risk assessment

Geodemographics can be a helpful tool for insurance companies finding that certain segments are particularly associated with areas that have high or low insurance claim rates or incidence of insurance fraud. Insurers have been able to build this knowledge into their models in order to better align pricing with risk.

The following contribution from Tony Lovick discusses the application of geodemographic segments to insurance pricing, as a readily available information source about consumers and their home neighbourhoods. Tony is an actuary with more than 20 years' experience in general insurance.

Application of geodemographics in insurance pricing

Tony Lovick, *Love Actuarially*

Most of us are familiar with the quote page on a comparison site that helps us to buy motor and household insurance each year. One of the reasons it feels so familiar is that the list of questions and the options in the drop-down for each question all appear the same wherever you go. This is because the comparison economy of scale is driven by the ability of every insurer to be compatible with a standard list of questions, and the Association of British Insurers (ABI) has facilitated that agreement. So it may appear that there is great inertia to maintain the status quo, giving the impression that there is little innovation or change in the way that insurers use this information year by year.

However, this impression could not be much further from the truth. Since the advent of computerized data records and the continual advances in computing power, insurers have been in an epic competitive battle to find any and every snippet of information they can usefully glean from the available data in order to edge ahead of their rivals. They also engage in a continual search for new sources of data to hook on to the quote that can be used for this purpose.

There are three perspectives to the risk that need to be considered – the person, place and insured object (vehicle, house and so on) – and useful external data can be attached to them all. Let's consider some examples of this and how it is done.

First of all the attachment process; in addition to the basic details of the risk, the quote form has some useful hooks that can be used to link external data. Notice that it asks for date of birth rather than age: providing age is clearly useful as a proxy for risk but providing date of birth allows identity checking. Then notice that the address fields are assisted by a look-up, which Postcode Address File (PAF) matches by asking for the full postcode and offering a drop-down to select the house within it. The general area you live in affects the risk, but the full postcode allows geodemographics to be appended at a very fine level of granularity. For car insurance it then asks for a vehicle registration number to assist in looking up the make/model/year, but not just to assess the risk by the type and age of the car; this also allows the vehicle identity to be checked.

With the big data revolution, the rate at which large sets of transactional data can be collected and synthesized, to provide insights into customer behaviour and attitudes, dramatically shifts the capability for external sources to be leveraged in a new way. Focusing on the individual, their identity is checked to highlight false or fraudulent identities so that they can be rejected from the process. Prior known details of their address and quote history can be compared to the current set for inconsistencies. Previous claims records can be searched, as can previously quoted rates. In responding to the renewal process a new quote or a renewal query might take account of existing quotes given through other channels, and the propensity for the customer to accept or decline based on their activity. Their Facebook comments, if public, and Twitter postings could be scanned for positive or negative feedback about the company brand, affecting their attitude to the product.

Focusing on the address, a quote could consider the theft risk and the influence on that from published crime rates in the area, or other relevant data about the location. For example, factors could include whether there is off-street parking, urban population, or proximity to certain high-risk places of interest.

In addition to these risk-specific attributes for each type of policy cover, geodemographic data is able to provide insights about the type of population in the street, family composition, employment and housing ownership – and these and other factors can all be included in the mix. Some data providers produce segmentations and scores based on this underlying data and these can certainly be useful as a way of understanding groups of customers for marketing purposes, plus the scores can be an easy win for modellers looking to find some correlation to claims.

More than likely the risk-modelling team will be very familiar with these datasets and will pull through all the underlying data attributes and look at these too. Typically, modelling from an overarching score or segmentation will yield lower predictive power than using the data they were computed from. This is because any summary will tend to lose information and this is especially the case where the aggregation method is formed from arbitrary or modelled weights that are not correlated to the response variable of interest to the insurer. If the components and sources of geodemographic data remain constant, the modeller would quickly become familiar with these inputs and move to a maintenance strategy dominated by automation and recalibration of existing results.

As we have seen, increased computer storage capacity has transformed the ability to keep transaction-level data. These newer data sources are challenging the modeller in different ways. For example, suppose the movements of a mouse and sequence of completing a quote form were captured in addition to the field entries themselves. Each mouse move alone cannot be considered to be a factor or even contain much information. To be useful the stream of these points needs to be interpreted into a pattern, and the presence or absence of that pattern can then be used as a variable in the model.

Rather than maintaining the status quo, the speed of innovation in this area has been increasing, allowing the new data scientist career path to command a higher premium than in the past.

Media applications

The use of geodemographics for targeting local advertising media – such as press, radio and posters – goes back to the 1980s. The following contribution from Venkat Anumula discusses these applications. Venkat is data science manager for the media agency Manning Gottlieb OMD, where he provides actionable insights to answer clients' questions.

Use of geodemographics in media research

Venkat Anumula, *Manning Gottlieb OMD*

Introduction

Geodemographics are used in the media landscape to better define and quantify the target audience and look at ways to reach out to the defined audience. In the past, advertisers were content with targeting one of the demographic segments such as age, gender, ethnicity, etc. However, increased competition and demanding consumer attitudes mean that advertisers cannot be satisfied with targeting broad demographic segments alone. They need to gain a holistic understanding of their audience and their potential value as a customer. That is where geodemographic analysis can be very useful. The benefit of using geodemographics is that it is created using the traditional variables such as age, gender, ethnicity and so on, but incorporates many more lifestyle and behavioural aspects including employment, financial and life-stage information, from multiple surveys such as the UK Census.

Key questions it can answer

Customer acquisitions would be the key remit of most media agencies. So every marketing campaign has to be clear on the following:

1   Who is the target audience?

2   Where do they live?

3   What is the headroom for growth from an acquisition perspective?

4   What method of advertising would reach them best? That is, what media channels do they engage with, and that would allow them to receive the communication with an optimal media spend?

5   What are their interests/behaviours so that they can be sent relevant content using the best possible medium?

The last point is very important because media budgets are not unlimited and marketers/advertisers have to be accountable for every pound they spend.

What can geodemographic analysis accomplish?

1   Improve customer/target audience understanding: since the total UK population is broadly classified into predefined segments, the segmentation is consistent across the UK, so it is possible to compare regions based on the geodemographic population profile.

2   Better communication with intended audience: we can overlay available media channels (ie TV, radio, press, digital, outdoor) on to customer hotspots. This feeds into media plans at a local/regional level.

3   Identify customer hotspots for acquisition campaigns: after a geodemographic profile is identified, we can locate regions that have a similar audience make-up. It is also possible to look at the audience market potential to gauge if it is worthwhile to increase media activity, thereby optimizing conversions.

4   Visual outputs: it is possible to output maps showing the catchment regions, their geodemographic composition and the media channels available to those regions. We can also overlay competitor presence or key locations of interest. This provides a visual and highly effective way of presenting data.

The process

Usually a postcode in the UK is uniquely identified with a geodemographic segment at the most granular geographic level, hence it is the starting point to carry out customer profiling, catchment definition, sales territory analysis (using nearest distance between customer and store), mapping and visualization.

So, given the list of existing customer postcodes, it is then possible to identify their geodemographic profile. Since we know the average penetration of various geodemographic segments in the UK, we next identify the over-indexing types against the total UK population. This information can be used to identify regions elsewhere that share a similar audience profile. Media information such as available TV stations, press, radio and out of home could be overlaid to aid media planning and strategy.

Furthermore, it is possible to profile a catchment area around a store using its postcode. Predefined drive times to store (30 minutes, 40 minutes, etc) can be overlaid to profile the population within the catchment. Competitor store presence can also be factored into this analysis. This feeds into creating benchmarks for store performance and will also contribute to local/regional media strategy.

CASE STUDY   Automotive client

Objective

An automotive client wanted to promote an eco-friendly car launch by focusing on particular areas of the UK rather than a blanket national approach.

Methodology

Using customer data from pre-orders, and leads for similar cars within their range, a geodemographic profile was built to understand which consumers were most likely to be interested in the new model. The dominant profile (indexing above the UK average) was then used to identify regional hotspots where the audience lived. In order to fine-tune regional selection, geographical information about eco-friendly car buyers was overlaid to uncover the most appropriate regions in which to run activity. To ensure that buying an eco-friendly car from the automotive client would be practical for potential customers in the identified regions, electric charging points and client dealerships were overlaid as well. For each region, TV, local radio stations and local press titles were identified.

Result

Five key regions across the UK were defined to form the focus of the car launch activity. Within each region, micro-targeting areas were then defined at postal sector level to inform direct communications activity. The overall campaign planning was tightly targeted regionally in order to minimize media wastage, thus optimizing media return on investment.

CASE STUDY   Telecommunications client

Objective

A telecommunications client wanted to grow their number of subscriptions from the student community by driving student footfall into the client's stores. They wanted to achieve this by targeting those people who live in residential areas.

Methodology

A target audience based on geodemographic segments was first identified as a proxy definition for students in general in the UK. Important university locations/campuses were looked at and the catchment area around them was profiled for the target audience. Next, the client's physical store locations were overlaid and their catchment area was mapped. This exercise helped to identify hotspot regions where there were the highest penetrations of students within the drive time of the client stores. The outputs formed the basis for local media planning around university campuses, student life areas and the client's physical stores.

Each physical store was ranked based on the potential to acquire new students as customers. The list of postal sectors to carry out the media activity was handed over to the relevant media-planning team.

Result

For each store, media investment was proportionally allocated as per the customer acquisition potential to account for local media activity. After the campaign went live, there was a significant increase in the customer subscriptions year on year from the student community.

Market research applications

The market research industry employs geodemographics at the data-collection stage, as a tool for selecting and controlling samples. The following contribution by Mark Watson, of Bluewave Geographics, explains how researchers do this, and illustrates the process with a case study example. Mark is an expert in creating sampling frames and carrying out geographical analysis for market research agencies.

Applications to market research

Mark Watson, *Bluewave Geographics*

The main ways in which geodemographic classifications are being used in the market research industry are:

- profiling;
- survey sampling;
- survey weighting;
- post-survey analysis.

Each of these will now be discussed, followed by a case study showing the use of a geodemographic classification in the selection of a survey sample.

Profiling

In order to make accurate inferences regarding a survey population from a market research survey sample, it is important that the sample is representative of the survey population. Whilst the level of confidence associated with survey results can be improved by an increase in the sample size, the results will not be reliable if there is inherent bias in the sample.

Survey area profiles

When conducting surveys it is first important to understand the characteristics of the survey population in terms of geography and demography. Using maps and profile reports, it is possible to build up a picture of the survey population and survey area.

As well as building up a demographic profile (for example from census data), a geodemographic classification may be used to profile and gain a greater understanding of the survey population. This profile is defined by the proportions of the survey population broken down by geodemographic types.

Once the geodemographic profile of the survey population has been built it can be used to help check the representativeness of any subsequently selected survey samples.

Sample profiles

As long as the postcode or census output area (OA) of each survey respondent is known, a profile can be built based on the geodemographic classification, defined by the numbers of survey respondents within each geodemographic type.

Survey sampling

Market research surveys are usually designed so as to be representative of the survey population in terms of both geography and demography. When selecting sample respondents, or sample points (or areas) in the case of a multi-stage sample, a method of sampling is often used whereby the full set of sampling units (known as the universe) is first sorted into a specific order before the sample is systematically selected using a random-start-and-fixed-interval technique. This technique is known as implicit stratification and is used by survey designers to help ensure that the sample is representative of the survey population in terms of the sortation variables. This technique will usually also reduce sampling errors.

The sampling units would typically be sorted first by a geographic indicator, and then within this by a demographic or socio-economic measure. For example, in the case study below the units are OAs and are sorted prior to selection by a geodemographic classification within each local authority.

Providing that the geographical units used in the first level of sortation are large enough to contain several sampled units, then the profile of the resulting sample should closely match that of the survey population in terms of both geography and the geodemographic classification used.

In the case of some surveys (such as tracker surveys) that are repeated over time in a series of waves, a geodemographic classification may be used to 'match' the samples wave by wave. This will help to reduce sampling errors when monitoring trends over time.

Sometimes, when conducting face-to-face surveys, the geographical area (or sample point) that an interviewer is working within may need to be replaced. For example, the quota may be unachievable or perhaps the area is deemed to be unsuitable. In such cases, to maintain the balance of the sample, a replacement area may be selected that is of the same geodemographic type.

Geodemographic classifications are also sometimes employed to target specific minority groups, for example users of a product. By honing in on geographical areas more likely to contain members of the target population, the number of screening interviews required to reach the minority group can be greatly reduced. Minority surveys on demographic groups, such as the elderly or mothers with young children, may alternatively be targeted with the aid of census data.

Not all surveys are designed to be representative of the survey population. In some cases surveys may deliberately over-represent particular geodemographic types in order to ensure that there are sufficient subsample sizes for meaningful comparisons to be made between these types. Weights would then need to be applied to the subsamples to correct for this when calculating overall survey estimates.

Survey weighting

Once the survey has been completed a profile can be built up of the respondents, in terms of geographical, demographic and socio-economic characteristics. This profile can then be compared with the profile of the survey population.

If there are significant differences between the profiles of the sample and the survey population then a weight can be applied to each respondent to help remove any measurable bias. For instance, if 20 per cent of the survey population falls into a particular geodemographic type, but this type makes up only 10 per cent of the sample, then a weight of two may be applied to each respondent in this type.

If the geodemographic classification is used in the sample design, and the survey is well designed and implemented, then geodemographic weighting should either not be necessary or the weights should be close to one.

Post-survey analysis

Once a survey has been completed a geodemographic code can be appended by postcode to each respondent record. It is then possible to analyse each survey response by neighbourhood type.

Geodemographic classifications are also used in more complex post-survey analysis such as clustering and segmentation, for example to profile and interpret a new set of segments.

CASE STUDY   A residents' survey in the county of Norfolk

The requirement is to select a random sample of individuals for a residents' satisfaction survey in the county of Norfolk.

Area profile

Using the Office for National Statistics's 2011 area classification for output areas (2011 OAC), a profile report can be drawn up comparing the profile of the survey area with that of Britain. This is shown in Table 7.1.

TABLE 7.1   Profile of Norfolk area

2011 OAC Super-groups	Survey Area %	Great Britain %	Index*	0 50 100 150 200
Rural Residents	38.8	11.9	326	
Cosmopolitans	3.0	5.2	57	
Ethnicity Central	0.5	5.6	8	
Multicultural Metropolitans	2.3	11.4	20	
Urbanites	14.7	18.2	81	
Suburbanites	17.5	19.6	89	
Constrained City Dwellers	10.0	9.7	103	
Hard-Pressed Living	13.3	18.4	72	

*The index is a comparison between the survey area % and the Great Britain %

Sample

As the survey is to be conducted face-to-face it is decided that the sample will be clustered in order to reduce fieldwork costs. A random-location quota sample will be used, in which 10 people will be selected for interview from within each of 100 sample points. Each sample point will be an OA.

The census OAs within the survey area are first sorted by geographical areas, and then within each of these units by the 2011 OAC code. In order for the second-level sortation to be effective it is important that each area contains several sample points. For example, if each area contained no more than one sample point, then the secondary sortation would be redundant.

Local authority districts are selected as geographical units for the first level of sortation. There are seven districts within the county, with the smallest accounting for 11 per cent of the survey population. So as each district will contain at least 11

FIGURE 7.1 Dispersion of sample across Norfolk by 2011 OAC

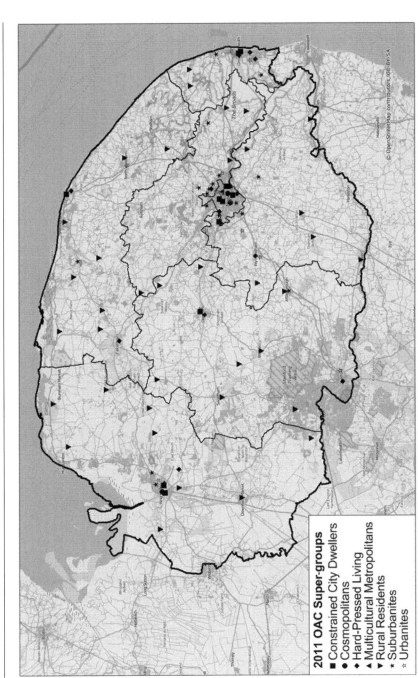

2011 OAC Super-groups
■ Constrained City Dwellers
● Cosmopolitans
◆ Hard-Pressed Living
▲ Multicultural Metropolitans
▼ Rural Residents
★ Suburbanites
☆ Urbanites

of the 100 sample points, the sampling units are sorted by 2011 OAC code within local authority district.

A systematic random-start-and-fixed-interval sampling technique is then used to select the 100 first-stage sampling units. Figure 7.1 shows the geographical dispersion of the sample points across the survey area, symbolized by the 2011 OAC classification.

Sample profile

Table 7.2 shows the 2011 OAC profile of the sample respondents in comparison with that of the survey area.

TABLE 7.2   Norfolk sample

2011 OAC Super-groups	Sample %	Survey Area %	Index*	0 50 100 150 200
Rural Residents	40.0	38.8	103	
Cosmopolitans	3.0	3.0	101	
Ethnicity Central	0.0	0.5	0	
Multicultural Metropolitans	2.0	2.3	87	
Urbanites	14.0	14.7	95	
Suburbanites	17.0	17.5	97	
Constrained City Dwellers	11.0	10.0	110	
Hard-Pressed Living	13.0	13.3	98	

*The index is a comparison between the sample % and the survey area %

Weighting

Due to the use of 2011 OAC in the sample design, the sample profile is a good match to the profile of the survey area. No geodemographic weighting is considered necessary in this case.

In summary

Geodemographic classifications are widely used in the market research industry. The applications of these classifications help researchers to gain a greater understanding of their survey population and provide considerable benefits in sample control, representativeness and accuracy. In addition, their use in post-survey analysis allows researchers to derive greater insight from survey results.

Public-sector applications

Public-sector organizations make use of geodemographics to help target services and communications for citizens in order to reduce wastage and improve efficiency. The same analytical techniques of profiling and mapping are employed as those we have seen used in the private sector. The following case study shows how one local authority has used these techniques.

CASE STUDY   Newport City Council

Challenge

The economic pressures facing all local authorities across the UK have been well publicized. At the time of writing Newport City Council, having achieved £50 million of efficiencies through its change programme over the last three years, is facing a further reduction of 13 per cent on its current budget, which will require additional efficiencies of £34 million by 2017. To meet the challenges the council needed to uncover new ways to deliver transformation and efficiencies beyond traditional methods.

Typically, public services have adopted a 'one size fits all' delivery model but the council recognized that if it could tailor service delivery according to customer demand and preferences, and drive online transactions, efficiencies would follow.

Solution

The council has developed a resilient delivery framework that is driven around efficiencies while ensuring citizens continue to get the services they need. Investment and improvements in other areas have taken place as a result of the efficiency programme. Within this framework sits Experian's Mosaic Public Sector customer segmentation, allowing the council to better understand its customers and build a detailed picture of its citizens' needs, preferences, behaviour patterns and trends.

Combining the insight provided through Mosaic with the council's own technical architecture – all of which is linked to a unique property reference identifier – the council could embed online services and subsequently improve the integration between customer relationship management systems and back-office applications.

The council is using this customer insight to develop a series of targeted projects in a number of areas. In 2013, the council launched its 'save time, go online' campaign to increase and improve channel-shift and speed up the return on investment from moving its citizen from more traditional face-to-face and telephone services to online self-service transactions. Forming part of its customer services strategy, the council wanted to increase choice for citizens and introduce more self-service opportunities.

According to Socitm (Society of Information Technology Management) figures, it costs £8.62 to facilitate each face-to-face transaction, £2.83 per telephone transaction and 15p per online transaction.

Result

As a result, in just under 11 months online transactions have increased from 24 per cent to 42 per cent. This will help to achieve savings of more than £500,000 over five years while offering an enhanced, 24/7 service for citizens. For customers who still require face-to-face interaction, the council has opened an award-winning multi-agency Information Station in the city centre.

The council is also revolutionizing the way it interacts with its citizens through customer relationship management by appending key customer insight characteristic information, including self-service and information channel preferences and use of council services in general. This has enabled the council to target specific profile household groups and subsequently to 'cross-sell' where and when appropriate, based on needs, wants and behaviour.

Applications in academia

From the cases outlined in this chapter, the reader could be forgiven for forming the impression that geodemographics is a quantitative analysis technique, of primary benefit for segmentation and targeting. We close this chapter with a case study which demonstrates that the applications are wider than one might imagine, extending in this case to social science research. The contributor, Tim Butler, is a former professor of human geography and of sociology; he has spent much of his career undertaking research into topics such as the regeneration of East London, education in the city, social mixing and gentrification.

The use of Mosaic in academic social science

Tim Butler, *King's College London*

Mike Savage and Roger Burrows in an important intervention in 2007 drew the attention of the sociological research community to the parallel world of what they termed 'commercial sociology'. Savage and Burrows (2007) pointed to some of the ways in which commercial sociology (broadly those drawing on loyalty cards and other forms of transactional data to increase the market penetration and ultimately the profitability of the commercial sector) worked to overcome the shortcomings of academic sociology in particular and social science more generally. They noted that academic social science, since its widespread introduction into the university curriculum in the post-war period, had first relied on the social survey (designed as Lazarsfeld and Rosenberg (1955) once noted to be equally at home on any topic, from understanding states of consciousness in a revolutionary situation to preferences for instant coffee) and then subsequently on the in-depth semi-structured interview. The former, whilst good at generating numbers, tended to be rather less good at generating answers to the 'why' question. This led in the 1980s to the widespread adoption of the so-called 'qualitative' approach, which of course might be very good at explaining why a particular respondent understood the world in a particular way but suffered from a lack of representativeness to a wider population.

The 'sample survey' was of course one of the genius technological inventions of the 20th century that enabled, assuming the sample was generated randomly from an appropriate population, social scientists to make accurate predictions with a known sampling error for that population. As indicated, however, the problem was that the stuff of such surveys was not usually very good at explaining *why* the population did what they did. The survey also tended to be 'one shot' and thus could not explain change over time.

Savage and Burrows therefore claimed that one of the problems of academic social science was that it relied over-heavily on inference of one sort or another, and they drew the attention of the academic community to the use by 'commercial sociologists' of transactional data. Transactional data has tended to record the way in which a whole population ('the universe') actually behaves rather than what they have told the investigator they might do. In other words, commercial sociology tended to deal with 'outcomes' for the whole population rather than 'stated intentions' of a sample drawn with varying degrees of rigour.

Geodemographic profiling, as has been made abundantly clear in the foregoing pages of this volume, is based around the collation and analysis of often literally hundreds of pieces of transactional data that record our lives – whether it be Tesco Clubcard, airline Frequent Flyer schemes or social-security payments. The advantages and disadvantages of this approach versus that of more conventional social sciences have been well rehearsed – notably in a follow-up (see *Sociology*, volume 43, issue 1) to the original Savage and Burrows article in *Sociology*, in which the latter's argument was taken to task by a series of authors as well as being trenchantly defended by the 'father' of geodemographics Richard Webber (2009).

The academic critique of the use of systems such as geodemographics could be seen as one of taste and culture but more substantially it focuses around two key weaknesses: the failure to engage with theoretical constructs and concepts that grapple with the enduring schisms in society and, relatedly, its lack of what can be seen as anchor concepts that endure across time and space. Key amongst these is social class, which works as both a theoretical and empirical concept in understanding social change and allows us to understand how society has changed over time. Commercial sociology in general, and geodemographics in particular, is more concerned, for understandable reasons, with how people consume in the here and now and is not overly bothered about how that might change or even, indeed, why different groups have different patterns of consumption. At one level, therefore, there is an entrenched antipathy between commercial and academic sociology. However, I am by no means the only academic researcher who has found that geodemographics, despite its shortcomings, has a lot that it can offer to an academic trying to understand social class change and emerging forms of inequality.

In the remainder of this short contribution, I give an account of how I have been able to use geodemographics in what is fundamentally a qualitative-based approach. Specifically, I have used Mosaic, probably the best known and most commercially successful of the geodemographic-based commercial applications.

I first came across Mosaic when I shared a platform at an ESRC-sponsored event on neighbourhoods in Bristol in 2004 with Richard Webber. We were both struck by the similarity of our findings about the new middle classes in London. I was talking about my recent research on an ESRC-sponsored study of the new middle classes and gentrification in London – published in Butler and Robson (2003) *London Calling* – in which I talked about not only how this metropolitan middle class differed from the non-metropolitan middle classes living in the suburbs, peri-urban housing developments, other cities and towns and rural areas but also how it differed across different areas of London. I argued – following the French sociologist Pierre Bourdieu – that there was a 'metropolitan habitus' that structured, and was in turn itself structured, socially *and* spatially by the experience of those who lived there. Put another way, certain areas appealed to certain groups (liberal-left intelligentsia, city technocrats, 'Hooray Henrys') and the act of living there confirmed these behaviours and attitudes. Essentially, the middle classes felt comfortable in these spaces. Richard was immediately struck not only by how this reflected the base assumption that inspired him to build Acorn, and more recently Mosaic, of how 'birds of a feather flock together' but, more practically, about how some of my descriptions of life in my chosen locations in inner London's gentrification belt accorded with some of his key groups (for example *Liberal Intelligence* and *New Urban Colonialists*). In a subsequent piece of collaborative work, Richard analysed the postcodes of my respondents and it was striking to both of us how accurately the Mosaic narrative descriptors matched the accounts of the respondents' lives from the quantitative and qualitative data they had provided to us. In particular, I was struck by the way in which Mosaic appeared able descriptively to bridge this divide and synthesize the often subtle distinctions that middle-class people in London drew between themselves and other groups, on the basis of their family and higher educational backgrounds and their broadly cultural and (small p) political preferences.

In the next large project that I undertook with my colleague Chris Hamnett, which was also initially about gentrification in East London but rapidly refocused on the part played by ethnicity and, among its new middle-class inhabitants the experience of schooling their children, we also drew on some of the insights of Mosaic. First, it aided us in choosing our research areas, which spanned across East London from the rather traditionally upper-middle-class gentrified enclave of Victoria Park to the outer former whitelands of Redbridge bordering the M25. The latter is now a destination of choice for aspirational British Asians from inner London. The findings of this study are written up in Butler and Hamnett (2011) *Ethnicity, Class and Aspiration: Understanding London's new East End*. We went further, however, than simply using Mosaic to scope out research areas for our survey work when, as indicated above, education and specifically the choice of school emerged as a key issue with our respondents and subsequently a major research focus. In a piece of multi-methods research, we brought together our qualitative interviews with 100 respondents drawn from a face-to-face survey with over 300 respondents in their own homes and an intensive analysis of the Pupil Level Annual School Census and National Pupil Database (PLASC/NPD). We undertook this in association with Richard Webber (Webber and Butler, 2007; Butler and Hamnett, 2007) in order to see what effect class and ethnicity were having on school attainment. One well-known problem is that PLASC does not record class background and as a social indicator records only being in receipt of free school meals (FSM). FSM is in effect an indicator of social deprivation and it is not clear even if those in receipt of FSM are representative of all those entitled to

FSM. By using Mosaic, and having been given confidential access to home postcodes from PLASC, we were able to apply the Mosaic classifications to show that home background and ethnicity both affected performance at GCSE, with the biggest effect being that of home background both positively and negatively, whilst ethnicity (and gender) also had similar but lesser effects. In a further paper, Hamnett, Ramsden and Butler (2007), we were able to show that the socially advantaged were the greatest choosers of high-achieving schools, either by living near them or else by ensuring that their children entered them even when they did not live in the catchment area (this was before the system was dramatically tightened up in 2006). It is not my purpose here to describe or discuss the findings of this research but to indicate that geodemographics made possible a form of analysis that was not possible using the data we had from PLASC or the normal tools provided by social science research methodology.

In this short contribution, I have tried to show that despite the fact that whilst geodemographics might be seen to have a natural home in so-called commercial sociology with its concerns with market segmentation and profit maximization, it can be used successfully in middle-range theoretical work such as that on gentrification and school choice.[1]

Conclusion

In this chapter, we have seen that geodemographic information can be applied in a variety of ways, tailored to the needs of different industry sectors and fields of study. All of these applications focus on gaining a better understanding of either people or local areas, and then using that knowledge to target services, business operations or research more effectively.

Note

1 I am aware that Mosaic has successfully been used to target social policy effectively (indeed its origins lie in designating the Educational Priority Areas in the 1970s) and also in getting out the party political vote (Barack Obama was the most successful candidate in doing this), but it is ironic that what started as a methodology for identifying the needy and directing scarce resources towards them has now become a system best known for filtering them out.

CHOOSING A GEODEMOGRAPHIC CLASSIFICATION

08

Introduction

As we saw in Chapter 4, eight general-purpose neighbourhood classifications are currently available in the UK. In addition to these, there are 12 market-specific discriminators and (as we saw in Chapter 5) 14 household and individual segmentations, along with other discriminators. Therefore, how should the new user decide on a classification system from some 40 products on offer in the marketplace?

The aim of this chapter is to provide some tips and guidance on selecting the most useful discriminator for your market and business application, by using an objective method of comparing the different systems. The following topics will be covered:

- Suggested steps for choosing a discriminator.
- Issues that often arise when comparing geodemographic products.
- Some widely used approaches for measuring the level of discrimination.
- A case-study example comparing geodemographic classifications.

Steps in the selection process

In principle, the process of choosing a geodemographic discriminator should be no different from the selection of any IT product or system. It should follow a series of logical steps, as illustrated below – the actual steps and terminology may be adapted to conform and comply with the procurement approach followed in your organization.

Step 1: Define the main purpose(s) of the discriminator – what will be its principal use(s) in your organization?
Defining the principal uses as fully as possible will help both your organization and your potential suppliers to focus on products and use cases that are relevant. For example, do you require a postcode classification for profiling enquirers and prospects, or a household-level segmentation for managing customers?

KEY POINT

It is better to identify just one or two definite uses, and specify them in detail, rather than produce a long list of possible use cases that are really only aspirations.

Step 2: Agree a set of selection criteria with the users.
For obvious reasons, the intended users should be involved throughout the selection process. They may have constraints or concerns that will need to be considered. For example, is there a limit to the number of segmentation categories that can be used? Is there a minimum viable match rate, for an individual-level discriminator, and how will unmatched records be handled?

Step 3: Identify products that appear to meet the selection criteria and issue a preliminary request for information (RFI) to their suppliers.
Based on step 2, produce a preliminary list of questions to obtain all of the essential details about each product and how it will satisfy your organizational needs. For assessment purposes, decide on how much weight to attach to the various questions and whether/how to score the responses.

Step 4: Assess responses to the RFI and decide on a shortlist of candidate products.
Remove those products that either do not satisfy your initial selection criteria or score very poorly across your set of questions.

Step 5: Send out a request for a proposal (RFP) to the shortlisted suppliers.
Based on the knowledge gained from the RFI stage, issue a detailed request for a costed proposal to supply the product, including licensing terms and arrangements for maintenance and updating. Design a comparative test of the products, based on sample data, and provide the dataset to suppliers submitting proposals.

Step 6: Carry out a detailed evaluation of the shortlisted products, including comparative analysis.
Obtain test results from each supplier, and analyse and draw conclusions about the usefulness of products when applied to your data. Compare suppliers' proposals and identify those for final consideration.

Step 7: Meet with representatives from suppliers for final assessments.
How well does each supplier seem to understand your business and what are their plans and vision for the future?

Step 8: Select preferred supplier and product, agree terms and formalize arrangements.
Arrange for your users to receive training and support from the supplier.

Although the aim of this process is to select a geodemographic discriminator, implicitly it also involves a choice of supplier – an important decision, as you will hopefully be working and sharing information with this company for many years to come. Therefore, the process should also be used to get to know new suppliers in your shortlist, start to understand their cultures, what they can provide in terms of future developments, and how well they would work with your organization. Table 8.1 provides a checklist of key questions that should be raised with candidate suppliers.

TABLE 8.1   Key questions for candidate suppliers

Support for your industry	Does the supplier have industry consultants with experience of working in your industry?
	Is the supplier's segmentation available on market-research sources used by your company?
	Does the supplier offer a market-specific segmentation for your industry?
Support and training	How will the supplier provide support and training for your users?
Supporting materials	What kinds of guides and documentation does the supplier provide in order to enable users to interpret their segmentation?
Software	Does the supplier provide software to operate their system? If so, what is its functionality and how useful is it?
Updates	How frequently are postcode changes supplied, and how will they be delivered to you?
	How often will the segmentation categories be updated? How realistic is the supplier's plan for keeping their classification up to date with changes to areas?
External links	Is the segmentation available on external databases used by your company? Can it be used to support online advertising?
Country coverage	Does the segmentation fully cover your company's trading area (eg UK)?
	Are international markets important to your business? If so, does the supplier provide equivalent systems for your key countries?
Supplier is regulated	Is the supplier regulated via membership of an appropriate professional or trade body?

For open-data products, such as OAC, the selection issues are naturally somewhat different. The product may be free to access and use; however, opportunity costs may be incurred if some of the commercial products would discriminate more strongly or better fit the needs of your business. Furthermore, there may be additional IT costs in deploying open products, such as maintenance of the postcode directory. Therefore a comparative evaluation of open and commercial discriminators is still recommended before making your choice.

Issues in comparing geodemographic products

Match rate

By 'match rate' we mean the proportion of your customer records that are recognized and matched to the supplier's database. For postcode-level discriminators, the match rate should be close to 100 per cent provided that customer postcodes are correctly formatted and reasonably up to date. However, for household and individual-level classifications, the match rate may be only 60–70 per cent, which would raise the question of how to treat the 30–40 per cent of records that are not matched.

If the main purpose of the classification is to segment and manage customers, using a household or individual-level product, then a high match rate would be essential. In this case, one of the components of evaluating candidate systems should be to supply samples of names and addresses for matching and data appending, comparing the resultant proportions of records matched and successfully classified.

Where the household/individual match rate is too low, a remedial action may be to apply the postcode-level segment to the unmatched records, but how well would this work in practice? Comparisons of profiles for matched versus unmatched records would help you to find out.

In the author's experience, customer records tend to be more successfully matched than prospects or enquirers, as customer name and address fields will invariably be more complete, so this should be borne in mind when designing such an evaluation. In other words, be sure to include separate samples of customers, enquirers and prospects in the analysis, if these groups are important to you.

> **KEY POINT**
>
> Measure relevant discriminatory power
>
> Depending on your principal use for the discriminator, it pays to design the evaluation in a relevant way that is indicative of this future use. For example, if the principal use were to help target e-mail campaigns, then the evaluation could analyse the geodemographic profile of respondents in comparison with the profile of the e-mail list, for each system being evaluated, in order to measure how well each system discriminates and identifies the target segments.

Allow for number of segments

As we saw in Table 4.2, geodemographic classifications differ from one another in terms of numbers of levels and segments at each level. A classifier with a large number of segments will automatically appear to discriminate more strongly than one with relatively few segments, due to random variation. This factor needs to be controlled for when comparing discriminators – this may be achieved by ranking the segments and combining them into fixed-size groups, such as deciles or quintiles, and comparing the discrimination shown by those groups.

Measuring discrimination

Geodemographic products may be assessed using the same techniques that a data-mining analyst would normally employ for evaluating targeting models, using graphical methods and numerical summaries.

Graphical methods

Continuing the example of profiling e-mail respondents by different classifications, a straightforward way to visualize their discriminatory power is to produce comparative gains and/or lift charts for them. These are two equivalent graphs that are often used for assessing data-mining models. They present essentially the same information in different ways, and the starting point for deriving either graph would be a series of customer profile reports (in spreadsheet format, for ease of calculation) for the various discriminators to be compared, each report profiling e-mail respondents against a base of e-mail contacts.

Gains charts

A typical gains chart, illustrated in Figure 8.1, plots the cumulative proportion of respondents (on the 'y' axis) versus the cumulative proportion of the base (on the 'x' axis), after ranking the geodemographic segments from highest profile index down to lowest. By ranking the segments in this way, we are assuming that the segments would be targeted in order of their indices, starting with the 'best' segment for reaching respondents, followed by the 'second best' and so on down to the 'worst' segment with the lowest profile index. This form of chart is also known as a receiver operating characteristic (ROC) curve or a Lorenz curve (after Max Lorenz who developed it in 1905 to measure inequality of the wealth distribution).

The diagonal line in the gains chart (the broken straight line) corresponds to equality in response captured versus proportion targeted, implying zero discriminatory power, and so represents a baseline for measuring each classification. In practice, the graph for each classifier will

FIGURE 8.1   Example gains chart

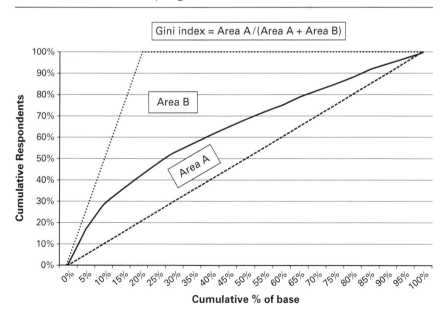

Gini index = Area A /(Area A + Area B)

Area B

Area A

Cumulative Respondents

Cumulative % of base

run above that diagonal line – the further from it, the greater the level of discrimination. The dotted line represents the gains graph for a perfect discriminator, which identifies 100 per cent of respondents in the top segments.

Lift charts

A lift chart, illustrated in Figure 8.2, is an alternative visualization, showing the cumulative lift – an index of the improvement in targeting compared against a random selection – with increasing proportion of the base contacted. Again, the segments are first sorted on their index values, from highest to lowest, before calculating the cumulative lift values.

FIGURE 8.2   Example lift chart

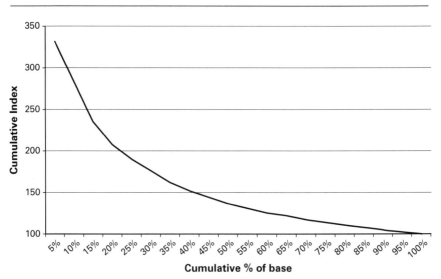

As for the gains chart, multiple discriminators may be compared in one lift chart. The line for each will start from the highest index and will descend, reaching 100 when 100 per cent of the base is contacted.

Numerical measures

Numerical measures aim to summarize the discriminatory power of each classifier, for ease of comparison. There are many possible measures and statistics that could be employed; however, several popular indicators may be derived from the gains and lift charts described above.

The first method is to select a proportion of the base that is likely to be targeted – say 50 per cent – and compare the classifiers, on their gains or lift, at that level of contacting. The outcome is a measure, for each product, of the response or lift it could deliver if applied to select 50 per cent from your base.

The second method measures the overall discriminatory power of each classifier, and focuses on the 'bow shaped' area of the gains chart, between the gains curve and the diagonal, shown as 'Area A' in Figure 8.1. For each classifier, this area may be calculated (in the customer profile spreadsheet) and compared with the total area (A + B) between the diagonal and the perfect discriminator (dotted line). The resultant ratio is known as the gini index and is often used as a summary measure of targeting efficiency.

Comparing geodemographic classifications

The following case study demonstrates the above methods of comparison in the context of market research data on ownership of digital tablets.

CASE STUDY   Comparison of geodemographic classifiers on tablet ownership using the British Population Survey

---

A marketer was planning to launch a chain of digital repair shops for tablets, and wished to select the most discriminatory neighbourhood classifier to help with site location and targeting local marketing activities. Therefore an evaluation was carried out across the available geodemographic products. The British Population Survey (BPS) was used to profile tablet ownership for each classifier, and a series of gains charts, lift charts and gini indices were produced. The analysis was based on 24 months of BPS data to 2014.

In total, nine neighbourhood classifiers were evaluated, all assignable at postcode level. These systems included both general-purpose and market-specific discriminators (due to the sensitivities of suppliers, the products have to be kept anonymous and so will be referred to as 'product A' and so on).

Figure 8.3 shows the values of the gini indices for the nine classifiers. It can be seen that, for tablet ownership, most of the products gave similar values of 15–19 per cent. This level of gini index suggests that the products would probably be useful targeting tools, ideally in combination with other selection criteria. Only product L gave a lower gini index of 8 per cent, implying that that this would be a significantly poorer tool for the tablet market.

FIGURE 8.3   Gini indices for nine post-2011 classifications, 24 months to 2014 (market: ownership of a tablet)

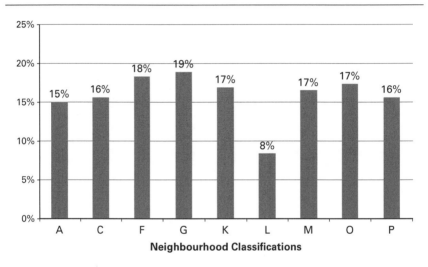

SOURCE: The British Population Survey

The marketer examined the gains charts for all nine classifiers; for illustrative purposes and clarity, Figure 8.4 shows the comparative gains chart for three of them – product G (with the highest gini index), product A (with a moderate gini index) and product L (lowest gini index).

From Figure 8.4, we can see that product G's gains chart appears marginally higher than product A's. For example, the marketer was able to read from the chart (or the underlying spreadsheet) that if a discriminator were being used to target 20 per cent of the population, then product G would reach 31 per cent of tablet owners, compared with 29 per cent for product A – a small but worthwhile difference, if costs and other factors were equal between the two systems. Product L displays a noticeably lower gains curve; using 'L' to target 20 per cent of the population would reach 25 per cent of tablet owners. Therefore product L would be an unattractive option, even it were available free of charge; in fact,

FIGURE 8.4   Gains chart for three post-2011 classifications, 24 months to 2014 (market: ownership of a tablet)

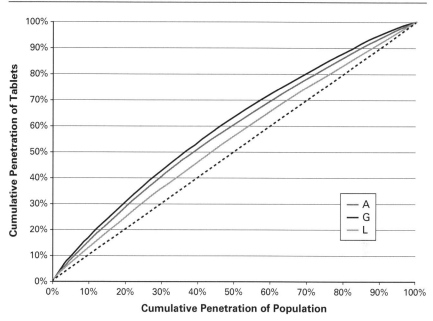

SOURCE: The British Population Survey

L was designed as a business-to-business tool and so its poor performance on tablets was hardly surprising.

Finally, the marketer also examined the set of lift charts; again, for illustration, Figure 8.5 shows the comparative lift charts for products G, A and L.

The lift chart illustrates the differences between products more clearly than the gains chart, particularly for the smallest, most discriminatory segments. The marketer can read the expected lift achieved by each discriminator, according to the proportion of population targeted. For example, if targeting 20 per cent as before, product G gives a lift of 152, ie 52 per cent better than a random selection. Next comes 'A' at 142, and 'L' is lowest at 124.

By evaluating the nine discriminators in this way, the marketer was able to select the most discriminatory products and assess their performance against their licensing costs.

FIGURE 8.5 Lift charts for three post-2011 classifications, 24 months to 2014 (market: ownership of a tablet)

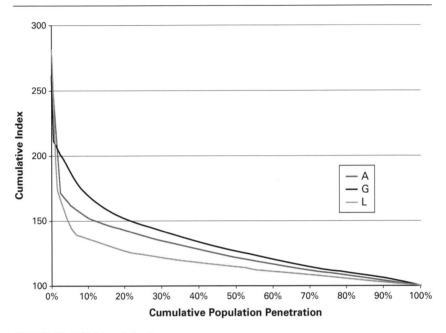

SOURCE: The British Population Survey

Other options

The British Population Survey may be applied for evaluating discriminators across a wide range of markets and lifestyles. Alternatively, one of the other large-scale syndicated surveys may be suitable (see Chapter 6, Table 6.4, for examples).

If the purpose of the geodemographic product is to enhance your in-house database, then another approach is to conduct the evaluation on your own customer data. This involves arranging for each supplier to apply its system to a sample of customer postcodes, and then carrying out the comparison for yourself.

Conclusion

In this chapter, we have seen that:

- Selecting a discriminator involves making a choice of both the supplier and the system – make sure that both are a likely fit for your current and future business needs.

- Geodemographic products should be compared in a way that is relevant to your main intended uses, taking account of differences in their numbers of segments.

- The graphical techniques of gains and lift charts are useful tools for comparing geodemographic discriminators.

- Discriminators may be compared using a suitable large-scale market research survey or by carrying out an evaluation on your customer data.

THE INTERNATIONAL PERSPECTIVE   09

Introduction

As we have seen, the geodemographics market in the UK is highly evolved, with a host of different classifications and discriminators available at neighbourhood, household and person level. However, geodemographics is not just a UK phenomenon – systems exist in many countries, enabling local companies and international businesses to benefit from segmenting those markets geographically.

The main aim of this chapter is to look at the international perspective. We start by discussing the main prerequisites for geodemographics to operate in a country. As census data plays a key role, we compare censuses and small-area output geographies in different countries, with the help of contributions from Peter Furness and David Martin. We then identify two sources of international classification systems, before closing the chapter with a case study on the status of geodemographics in the United States, by Kyle Poppie and Dave Miller.

Prerequisites for geodemographics to operate in a country

Going back to the 'geo' and 'demographics' components of geodemographics, two main data resources are required for this science to operate in a country – a national source of demographic data, at small-area level, and the ability to locate the data geographically 'on the ground'. We consider each of these prerequisites below.

Demographic data at small-area level

A source of census-type data is an essential prerequisite for geodemographics to operate and, at the same time, the data has to be available for small areas or zones across the entire country. In more developed countries, this joint requirement is satisfied by the census of population.

A country's census is a national tool, conducted by its government primarily in order to help plan delivery of services to citizens, allocate funding and quantify national issues. Therefore, as we shall see below, while some harmonization has been introduced by the United Nations (UN) and the European Union (EU), each census will include questions designed to examine the population in terms of the government's priorities. And while sample surveys can deliver that information at a national level, only a census can give the picture down to each local area.

In the UK, for example, immigration has been an important issue for some years, therefore the 2011 census included new questions designed to examine this, including national identity, passports held, length of time in the UK and how long staying. In Brazil, on the other hand, housing conditions and sanitation are issues, and so the country's 2010 census contained a number of questions on these topics – including a question on how each household disposes of its rubbish: collected, stored in dumpsters, burned, buried or thrown away.

Ability to locate data geographically

As discussed in earlier chapters, geodemographics entails locating segments geographically, therefore resources are required such as boundary files for census small areas – so that the segments can be mapped – and look-up tables linking small areas to consumer addresses.

Challenges in developed and developing countries

Developed countries are likely to satisfy both of these prerequisites, and so may well have geodemographic solutions built for them. Achieving consistency between country segmentations is another issue – as census topics are likely to differ, so are the neighbourhood classifications. This presents a challenge for international users, who would want to segment consumers in a consistent way across all of the countries they trade in.

Developing countries face a number of challenges for geomarketing, including the absence of one or both of the above prerequisites. Nicholas Allo (2014) discusses the issues in the context of geodemographic development in Nigeria.

Differences between national censuses

Given that a country's main source of demographic data is likely to come from a census, the following contribution from Peter Furness reviews the status of census taking around the world. Dr Furness is a consultant mathematician who throughout his career has specialized in the development and application of new analytical techniques, including geodemographics.

Census taking around the world – the topics being captured and the methods being followed

Peter Furness

In this section we examine the availability of census-based small-area population data from around the world. Our focus is mainly on the European Union but other countries, especially those in the United Nations Economic Commission for Europe (UNECE), are also mentioned. First, we look at why countries conduct a census, and then look at the principal methods of census taking and how this affects the topics under discussion.

Drivers for the census

National population censuses are the only statistical form of investigation that can produce an accurate quantitative picture of the population structure, households and socio-economic conditions of a country at a local level. Census taking was institutionalized in most European countries during the 19th century and the main principles of census taking were introduced and internationally acknowledged by 1846 – these being self-enumeration of the whole population with household and individual questionnaires based on scientific methods.

The 20th century saw steady growth in terms of the breadth and depth of topics covered by censuses. However, since the 1970s, increasing problems with response rates, and pressures on costs, have prompted some countries to look for alternatives to the traditional questionnaire-based data collection.

From its foundation the United Nations has recommended that national governments hold population censuses at the end or beginning of each decade (United Nations, 2008). The UN Statistics Commission estimates that 223 countries will have conducted a census during the period 2005–14 (UNSD, 2010), enumerating approximately 98 per cent of the world's population.

The United Nations recommendations also include census topics and these have been refined and extended for UNECE countries with lists of so-called core and non-core topics (UNECE, 2006). The EU has agreed on a minimum set of topic questions, based on UNECE recommendations, and these are enshrined in EU regulations (Eurostat, 2011). For a list of core and non-core topics see Table 9.1.

TABLE 9.1   List of proposed core and non-core topics for the 2010 round of population and housing censuses, CES countries

Core Topics	Non-core Topics
Population to be Enumerated	
Place of usual residence	
Total population (derived)	
Geographic Characteristics	
Locality (derived)	*Urban and rural area (derived)*
Location of place of work	Location of school, college or university
	Mode of transport to work
	Mode of transport to school, college or university
	Distance travelled to work and time taken
	Distance travelled to school, college or university and time taken
Demographic Characteristics	
Sex	De facto marital status
Age	Total number of children born alive
Legal marital status	Date(s) of legal marriage(s) of ever married women: (i) first marriage and (ii) current marriage
	Date(s) of the beginning of the consensual union(s) of women having ever been in consensual union: (i) first consensual union and (ii) current consensual union
Economic Characteristics	
Current activity status	Usual activity status
Occupation	Providers of unpaid services, volunteers
Industry (branch of economic activity)	Type of sector (institutional unit)
Status in employment	Informal employment
	Type of place of work
	Time usually worked
	Time related underemployment
	Duration of unemployment
	Number of persons working in the local unit of the establishment
	Main sources of livelihood
	Income
	Socio-economic groups (derived)

TABLE 9.1 *cont'd*

Core Topics	Non-core Topics
Educational Characteristics	
Educational attainment	Educational qualifications
	Field of study
	School attendance
	Literacy
	Computer literacy
International and Internal Migration	
Country/place of birth	Country of previous usual residence abroad
Country of citizenship	Total duration of residence in the country
Ever resided abroad and year of arrival in the country	Place of usual residence five years prior to the census
Previous place of usual residence and date of arrival in the current place	Reason for migration
	Country of birth of parents
	Citizenship acquisition
	Persons with foreign/national background (derived)
	Population groups relevant to international migration (derived)
	Population with refugee background (derived)
	Internally displaced persons (IDPs) (derived)
Ethno-cultural Characteristics	
	Ethnicity
	Language
	Religion
Disability	
	Disability status
Household and Family Characteristics	
Relationship between household members	*Same-sex partnerships (derived)*
Household status (derived)	*Extended family status (derived)*
Family status (derived)	*Type of reconstituted family (derived)*
Type of family nucleus (derived)	*Type of extended family (derived)*
Size of family nucleus (derived)	*Generational composition of private households (derived)*
Type of private household (derived)	Single or shared occupancy
Size of private household (derived)	Rent
Tenure status of households	Durable consumer goods possessed by the household

TABLE 9.1   *cont'd*

Core Topics	Non-core Topics
	Number of cars available for the use of the household
	Availability of car parking
	Telephone and internet connection
Agriculture	
	Own-account agricultural production (household level)
	Characteristics of all agricultural jobs during the last year (individual level)
Living Quarters, Dwellings and Housing Arrangements	
Housing arrangements	Availability and characteristics of secondary, seasonal and vacant dwellings
Type of living quarters	Occupancy by number of private households
Location of living quarters	Type of rooms
Occupancy status of conventional dwellings	Hot water
Type of ownership	Type of sewage disposal system
Number of occupants	Kitchen
Useful floor space and/or number of rooms of housing units	Cooking facilities
Density standard (derived)	Main type of energy used for heating
Water supply system	Electricity
Toilet facilities	Piped gas
Bathing facilities	Air-conditioning
Type of heating	Position of dwelling in the building
Dwellings by type of building	Accessibility to dwelling
Dwellings by period of construction	Lift
	Dwellings by number of floors in the building
	Dwellings by materials of which specific parts of the building are constructed
	Dwellings by state of repair of the buildings

NOTE: Derived variables are shown in italics
SOURCE: UNECE (2006)

The principal methods of census taking and their advantages/disadvantages

Census taking covers a wide spectrum of methods, ranging from the traditional questionnaire-based census through to those using administrative registers. In between are various combinations of the two, supplemented in some cases by various types of sample survey, either in conjunction with the census (sample enumeration) or as separate activities (household surveys) (Furness, 2004).

The traditional census

The traditional census – typified by the UK and many major economies including the United States, China and Japan – involves the distribution of census questionnaires that are then collected by some combination of enumerators, post-back, or web-based response.

The traditional census has several advantages. One is that the national statistical organization can manage the census to maximize the statistical potential of the data collected. Another is the comprehensive nature of the census in terms of topic coverage. Also, public awareness engenders a sense of national participation and leads to exceptionally high response rates unmatched by conventional surveys.

Cost is cited as the main disadvantage of the traditional census. Field operations and data processing are the major costs and these are being driven upwards by, amongst other things, the need to maintain response rates in the face of increasing mobility as well as reluctance on the part of the public to participate (Valente, 2010).

The infrequency of the traditional census is also a disadvantage. Starting from scratch every 10 years can lead to organizational headaches. Additionally, the census data, though current at the time of collection, will less accurately reflect local conditions after 5–10 years, and this can be a major problem for some users.

Finally, traditional censuses are finding it hard to grapple with certain sub-populations such as the single young mobile population, people living in ghettos or shanty towns and marginalized sections of the community.

Sample surveys

For cost reasons, many countries carry out a sample enumeration in conjunction with the census in order to collect more detailed information on a separate (longer) questionnaire, as in the United States and China. This is a cost-effective way to broaden the scope of the census. However, any sample survey has limited applicability for collecting statistics at small-area level, due to sample-size issues.

Some countries employ separate sample surveys in their census operations. These are the most flexible of the census data sources. In principle, almost any subject can be investigated. Countries that are using household surveys include France and Germany. In France, a unique 'rolling census' has been conducted since 2004 in which annual sample surveys are combined to cover 70 per cent of the French population over a five-year cycle (Valente, 2010).

While household surveys are not as expensive as full population censuses, they are costly to organize, particularly at the beginning when countries do not have an established programme of household research. They do not usually provide sample sizes sufficient for detailed local analysis and, like all surveys, they suffer from an increasing unwillingness of the public to participate – unless made compulsory.

Administrative registers

The third most important and commonly used data source is administrative registers. They have an important benefit in that the same information need not be collected more than once from the same individual for statistical purposes. Crucially, some form of civil register or maintained individual and household database is essential as a data spine against which other data sources can be matched (United Nations, 2008).

Costly large-scale field procedures are avoided and most effort is concentrated on improving the quality of the registers themselves and on using statistical methods to derive the required output. Some small-scale field procedures may still be required to validate administrative data and ensure that data quality reaches targeted levels.

Nordic countries have relied on register-based statistics for their data needs since the 1960s. Austria now also has a register-based census and several EU countries combine registers with either traditional enumeration (eg Spain, Czech Republic, Estonia, Italy, Latvia, Lithuania) or with sample surveys (eg Belgium, Netherlands, Slovenia). Some countries even use a combination of registers with both enumeration and sample surveys (eg Germany and Poland).

Among the disadvantages of administrative registers are that they are often limited in content and their uses may be restricted. Nor are they as adaptable as household surveys in terms of content or subject detail and are often incomplete, inconsistent or limited in coverage (especially in areas such as health, nutrition or household expenditures). Moreover, administrative records often focus on the individual, rather than the household or family, limiting the scope for social analysis.

A 2007 review by European statisticians highlights many of the issues relevant to register-based censuses, including difficulties with topics and the definitions of variables (UNECE, 2007). Two of the countries that moved to a fully register-based census in the 2011 round, Norway and Austria, have documented many of these challenges and show how the UNECE core topic requirements can be largely met, albeit with improvements to the registers and resorting to sample surveys for problematic core variables such as occupation (Statistics Norway, 2007; Lenk, 2008).

Other drawbacks to using administrative registers include concerns over privacy and the misuse of public-sector information.

Survey of methods used

In June 2009 the United Nations Statistics Division (UNSD) conducted a worldwide survey into census data-collection methods for the 2010–11 round of censuses. The results are shown in Table 9.2 for the 50 UNECE countries who responded to the survey, including all 27 EU member countries (UNECE Secretariat, 2010).

For the EU countries, only 41 per cent (11 countries) planned for a traditional census. The same percentage plan to use a combined approach with the remainder (four countries) conducting a purely register-based census. France is unique in conducting a rolling census.

For non-EU countries within UNECE, a much greater proportion are planning a traditional census (74 per cent), with four countries planning a combined approach and just one country (Norway) conducting a register-based census. The United States conducted a traditional census in 2010 but this will be integrated with yearly updates of individual characteristics, based on a large annual sample survey (The American Community Survey – see below).

Trends and future developments

The 2009 UNSD survey (UNECE Secretariat, 2010) also examined the changes in census methodology that had occurred between the 2000 and 2010 census rounds. The trend away from the traditional census is most marked in EU countries where seven out of 18 countries that conducted a traditional census in the 2000 round have conducted either a combined approach (Czech Republic, Estonia, Italy, Lithuania, Poland), a purely register-based approach (Austria) or a rolling census (France). See Table 9.3.

TABLE 9.2   Census type and date for 2010–11 round, UNECE countries

Country	EU	Census Type	Census Date
Albania		Traditional	4/1/2011
Armenia		Traditional	10/12/2011
Austria	Yes	Register-based	10/31/2011
Azerbaijan		Traditional	4/13/2011
Belarus		Traditional	10/14/2009
Belgium	Yes	Combined (registers and survey)	1/1/2011
Bosnia and Herzegovina		Traditional	3/31/2011
Bulgaria	Yes	Traditional	3/10/2011
Canada		Traditional	5/10/2011
Croatia		Traditional	3/31/2011
Cyprus	Yes	Traditional	10/1/2011
Czech Republic	Yes	Combined (registers and enumeration)	3/26/2011
Denmark	Yes	Register-based	1/1/2011
Estonia	Yes	Combined (registers and enumeration)	9/18/2011
Finland	Yes	Register-based	12/31/2010
France	Yes	Rolling census	1/1/2011
Georgia		Traditional	2012
Germany	Yes	Combined (registers and enum and survey)	5/9/2011
Greece	Yes	Traditional	3/16/2011
Hungary	Yes	Traditional	10/1/2011
Iceland		Combined (registers and survey)	not available
Israel		Combined (registers and survey)	12/27/2008
Ireland	Yes	Traditional	4/1/2011
Italy	Yes	Combined (registers and enumeration)	10/23/2011
Kazakhstan		Traditional	2/25/2009
Kyrgyzstan		Traditional	3/24/2009

TABLE 9.2 *cont'd*

Country	EU	Census Type	Census Date
Lithuania	Yes	Combined (registers and enumeration)	3/1/2011
Luxembourg	Yes	Traditional	2/1/2011
Malta	Yes	Traditional	11/1/2011
Montenegro		Traditional	3/31/2011
Netherlands	Yes	Combined (registers and survey)	1/1/2011
Norway		Register-based	11/19/2011
Poland	Yes	Combined (registers and enum and survey)	3/31/2011
Portugal	Yes	Traditional	3/1/2011
Republic of Moldova		Traditional	4/1/2012
Romania	Yes	Traditional	3/1/2011
Russian Federation		Traditional	10/14/2010
Serbia		Traditional	3/31/2011
Slovakia	Yes	Traditional	5/21/2011
Slovenia	Yes	Combined (registers and survey)	1/1/2011
Spain	Yes	Combined (registers and enumeration)	11/1/2011
Sweden	Yes	Register-based	12/31/2011
Switzerland		Combined (registers and survey)	12/31/2010
Tajikistan		Traditional	10/1/2010
The Former Yugoslav Republic of Macedonia		Traditional	3/31/2010
Turkey		Combined (registers and survey)	2011
Ukraine		Traditional	2011
United Kingdom	Yes	Traditional	3/27/2011
United States		Traditional enumeration with yearly updates	4/1/2010

SOURCE: UNECE–UNSD survey 2009

TABLE 9.3 Census type in 2000 and 2010 rounds, EU countries

		Census Type in 2010 Round				
		Traditional	Combined	Register-Based	Rolling	Total
Census Type in 2000 Round	Traditional	11	5	1	1	18
	Combined	0	5	0	0	5
	Register-based	0	0	2	0	2
	No census	0	1	1	0	2
	Total	11	11	4	1	27

SOURCE: UNECE–UNSD survey 2010

In 2014, the UK decided to carry out a more-or-less traditional census in 2021 along with further research into the use of administrative registers (see Chapter 11). The indications are that the trends reported in Table 9.3 will certainly continue over the next decade, at least as far as the EU is concerned (White, 2014).

This section is based on the author's contribution to an article published in *International Journal of Market Research* (Dugmore et al, 2011).

Differences in small-area output geographies between countries

Just as census methodology varies between countries, so does the approach taken for defining each country's geography of small areas for census output – potentially its units for geodemographic segmentation. The following contribution from David Martin, Professor of Geography at Southampton University, compares the small-area output geographies used by different countries. Professor Martin is an expert in geographical referencing and modelling small-area population data; he devised the current system of output areas used for the 2001 and 2011 censuses in England and Wales.

Comparison of small-area output geographies between countries

David Martin

Contemporary censuses, administrative and survey-based population data systems result in national statistical organizations (NSOs) holding databases in which records contain the social and economic characteristics of individuals and households. These are geographically referenced in a variety of ways but increasingly contain high-resolution grid coordinates, an address or other property-level reference. Often, there will also be geographical codes associated with the areas used for enumeration or survey administration, although these are not necessarily the spatial units for which statistical outputs are to be produced (United Nations, 2009). In most countries, the record-level data is treated as confidential and it is therefore necessary to aggregate it geographically to produce statistical outputs for small areas. Considerations regarding the design of these small areas are largely independent from the data collection strategy used, as once a high-resolution spatial reference is included in the population database it becomes relatively straightforward to aggregate to any desired output geography, whether the records are originally derived from linkage of administrative registers or a conventional census.

The definition of 'small' area in this context varies widely between countries: many of the policy-based motivations for collecting census-type data in the first place are best served by having statistics for very local areas that provide flexibility in further aggregation by users. There is thus a tension between protecting confidentiality and producing the most useful small-area outputs. In general, a hierarchical approach is adopted with less detailed statistics being released for the smallest areas, often with additional statistical devices such as minimum population size thresholds, perturbation or rounding of counts in order to further protect confidentiality. There is generally reluctance to produce the same statistics for multiple types of small areas due to the potential risk of disclosure by differencing: if statistics are published for two slightly different areas it is possible to work out the values for the small area of difference (Duke-Williams and Rees, 1998).

NSOs are therefore faced with many potential options for small-area output geography and it is hard to find much international standardization, each country having systems of internal boundaries that reflect its unique cultural and political history. Administrative or electoral divisions determine the structure of most census small areas. There is almost always a requirement to provide core statistics for regional and local government, typically the basis for service delivery, allocation of resources and political representation. These geographies have often been in use for many decades. Hierarchical systems are common, with larger units often reflecting historical municipality boundaries and the smallest units more likely to reflect modern statistical considerations in terms of population size, geographical extent or other formal design criteria. The availability of extensive national spatial data infrastructures and widespread use of geographical information systems (GIS) offers many candidate boundary features that could be used in defining such areas. While some countries produce statistics for very small areas, whose boundaries reflect the physical structure of streets and properties, others – particularly in Scandinavia – have adopted the strategy of generating at least a subset of outputs for regular geographical grids. Grid cells are of interest because they remain constant over time while all the alternatives are subject to revision, due to changes in administrative systems or the underlying population distribution. When boundaries are revised it becomes impossible to definitively assess population changes, due to the difficulty of disentangling the two sources of change. This is an instance of what Openshaw (1984) termed the 'modifiable areal unit problem' – that different boundaries will result in differing statistics, even from the same underlying population (discussed further in Chapter 6).

Clearly, it is not possible to discuss in detail here the small-area geographical data available for every country, so the following represents a selection of countries that illustrate some commonly encountered features of census small-area output geographies. Table 9.4 summarizes key characteristics, focusing primarily on countries for which the boundaries of the smallest areas may be freely downloaded.

Different parts of the UK conduct independent simultaneous censuses. In England, Wales and Northern Ireland, a new automated zone design procedure was introduced in 2001 to generate census output areas. The procedure sought to explicitly calculate the best trade-off between competing requirements. This design process indicates some of the many different criteria that might be important. Polygons were generated for each unit postcode, the smallest entity in the postal delivery coding system. These polygons were then automatically aggregated to produce the best combinations within larger wards and parishes, trading off population-size thresholds (100 persons and 40 households) and target sizes (125 households), shape constraints and a social homogeneity measure (Martin, 2002). Boundaries were constrained where possible to follow major roads and other topographic features. Broadly the same approach was applied in England and Wales in 2011, including creation of an alternative set of new units for the publication of workplace data, known as workplace zones (Martin, Cockings and Harfoot, 2013). Scotland, by contrast, has adopted its own approach to the design of output areas since 1981, using lower confidentiality thresholds and aiming to minimize change between censuses. Another automated solution, built around the street network (Fotheringham, Foley and Charlton, 2008) was introduced to generate small areas for Ireland's 2011 census. In each of these cases the smallest output areas in use were explicitly developed for the purpose of publishing census data, taking into account the census population counts.

While the England and Wales approach has focused on creating small areas with consistent population sizes at the cost of substantial change between censuses, the approach in France has been to use historically consistent communes, with the result that there are enormous variations in population size, from over two million in Paris to a few that are entirely unpopulated. France's transition to a rolling census methodology uses the communes as the basic sampling units, but with somewhat uneven coverage, as there is further subdivision of larger communes (only) into smaller statistical units. The 2011 censuses in Sweden and Switzerland were fully based on administrative registers (Ralphs and Tutton, 2011). This approach again offers the potential to produce any desired aggregation from individual records, but in practice has not led to the creation of smaller output areas. Within Europe, the Nomenclature of Territorial Units for Statistics (NUTS) scheme attempts to reconcile the many different national systems by matching each system on to the NUTS hierarchy, thus French departments are considered equivalent to UK local authorities (with some grouping) at NUTS level 2. Below the three NUTS levels, two levels of Local Administrative Units (LAUs) are similarly matched with, for example, French communes, Swedish municipalities and UK wards at LAU level 2. Census statistics are generally available for all countries down to at least LAU level 2, but the smallest output areas for the UK and Ireland are all more detailed than LAU2.

Australia's smallest units, known as mesh blocks, have a mean population size of only 62 but again form the lowest level in a hierarchy of geographical units with only basic counts published at the mesh block level and detailed statistical outputs for larger units (ABS, 2011). This system is comparable to that in New Zealand, where the units are also known as mesh blocks. It is a feature of countries containing very thinly settled regions that the 'smallest' areas may in fact need to cover huge geographical extents in order to contain sufficient population. In the United States, work to develop a street-based referencing system for digital mapping of the 1970 census was itself an important driver of modern GIS data structures (Peucker and Chrisman, 1975). This has developed into a hierarchy of census areas fundamentally based on aggregations of addresses into blocks, the smallest areas for statistical outputs. Census block boundaries in the United States and dissemination blocks in Canada (Statistics Canada, 2012) are defined in large part by street-block faces but with extensive use of other topographic features (United States Census Bureau, 1994), particularly in remote and rural areas.

TABLE 9.4 Key characteristics of small-area output geographies in different countries

Country/ Reference Year	Name of Smallest Output Area	Number of Units	Mean Population Size	Source of Freely Downloadable Digital Boundary Data (geographic area boundaries portal or home page), if available
England and Wales/2011	Output area	181,408	309	https://geoportal.statistics.gov.uk/geoportal/catalog/main/home.page
Scotland/2011	Output area	46,351	114	http://www.nrscotland.gov.uk/statistics-and-data/geography/our-products/census-datasets/2011-census
Northern Ireland/2011	Small area	4,537	399	http://www.nisra.gov.uk/geography/SmallAreas.htm
Ireland/2011	Small area	18,488	248	http://www.cso.ie/en/census/census2011boundaryfiles/
France	IRIS Statistical block group	50,100	Varies widely	N/A
Portugal/2011	Statistical subsection	177,893	59	http://mapas.ine.pt/download/index2011.phtml
USA/2014	Census block	11,078,297	28	https://www.census.gov/geo/maps-data/data/tiger.html
Canada/2011	Dissemination block	493,345	68	http://www12.statcan.gc.ca/census-recensement/2011/geo/bound-limit/bound-limit-2011-eng.cfm
Australia/2011	Mesh block	347,627	62	http://www.abs.gov.au/ausstats/abs@.nsf/mf/1209.0.55.002/
New Zealand/2013	Mesh block	46,621	91	http://www.stats.govt.nz/browse_for_stats/people_and_communities/Geographic-areas/digital-boundary-files.aspx

The relationship between national statistical organizations (NSOs) and national mapping agencies (NMAs) varies, with census mapping depending to a greater or lesser extent on spatial data infrastructures provided for other purposes by NMAs. An increasingly common feature of the small-area-boundary datasets has been the move towards free and open publication, evidenced by the examples in Table 9.4, as open data initiatives have gained ground. It is increasingly common to be able to undertake basic mapping online as well as to directly download the boundaries as, for example, Statistics Portugal's 2011 census mapping interface (**http://mapas.ine.pt/map.phtml**) or US Census Bureau maps and data pages (**https://www.census.gov/geo/maps-data/**). The boundary datasets are frequently also accompanied by ancillary look-up tables that define the official association between current and past census small areas, and other spatial divisions such as electoral or postal geography.

It is apparent from this brief review that despite very diverse population densities, physical environments and administrative systems, comparable hierarchical small-area output geographies are encountered internationally. Perhaps the most important aspects to be considered for spatial analysis are the nationally specific scale of the smallest areas and the statistical detail considered acceptable for publication at this scale, which displays wide variation.

International classification systems

Once geodemographics had proven itself as a powerful segmentation tool in the UK, suppliers started to look beyond its shores for other countries that would support having their own neighbourhood discriminators. In the year that Experian launched their UK classification, 1985, they followed it with a segmentation for France. This was followed one year later by a system for the Netherlands.

As of 2015, two of the UK suppliers have taken a lead in building equivalent systems in large numbers of other countries – these are Callcredit and Experian. Their coverage of countries is broadly similar, mainly covering the developed countries in Western and Eastern Europe, North America and the Asia/Pacific region. Their international classifications are overviewed briefly below.

Callcredit's international CAMEO classifications

Callcredit has developed CAMEO classifications in 34 countries apart from the UK, selected in part by ranking all countries on financial criteria to select initial markets, adding countries of strategic importance and considering data availability.

Where census data does not go down to neighbourhood level, Callcredit will sometimes split data for larger areas into smaller units, based on known characteristics, and then build a small-area classification.

An international CAMEO code has been created, based on life stage and affluence, which assigns the separate CAMEO codes for each country into a consistent set of groups.

Experian's international Mosaic classifications

Experian has built Mosaic classifications in 25 countries beyond the UK. In countries where census output is only released for larger areas, Experian has developed a methodology that models census results down to one-kilometre grid squares with the aid of other data, including satellite imaging. A Mosaic classification is then developed using the modelled data for grid squares.

Experian has also created an 'umbrella' segmentation that assigns the Mosaic types for each country into consistent groups. A set of demographic variables is used to allocate each country's Mosaic types to a Mosaic Gold group.

Obtaining international data

It is easier to access international data nowadays, with census output increasingly being made available online. Therefore, if you are interested in obtaining data for a particular country, an online search for the NSO or census agency is a good place to start.

The Geodemographics Knowledge Base (available at: **http://www. geodemographics.org.uk/**) includes a Europe section that contains links to sources of international statistics, including the NSOs for many European countries.

Country case study – the United States

The geodemographics market in the United States dates from the late 1970s and so is of a similar age to the UK market. The first US classification system was PRIZM ('Potential Rating Index for ZIP Markets') developed by Claritas using 1970 census data.

The following case study, conducted by Kyle Poppie and Dave Miller, overviews the US market and the issues it is facing. The same issues are likely to apply soon in the UK, if they are not already here. Kyle Poppie is director of data science for Nielsen in the United States, and is experienced in analysis of purchasing, viewing and demographic data. David Miller is vice president of methods in measurement science in the Nielsen Centre of Innovation. Dave is an expert in segmentation and has been working with demographic and lifestyle data for more than 30 years.

CASE STUDY   US market for geodemographics

Kyle Poppie *and* **Dave Miller**

If there has been one characteristic that defines geodemographic systems over their history it has been the progression towards classifying smaller and smaller levels of geographies. Most systems now function at the household level, with some even operating at the person level. That is not to say that this is always the preferred client solution, as many clients still use postal level such as ZIP+4 (see Table 9.5 for commonly used geographic concepts and their approximate household counts in the United States).[1]

The progression towards smaller geographies has been driven by the development of large-scale databases containing characteristics and behaviours at these levels. At this point nearly all US systems in common use are available at a household or person level, as seen in Table 9.6.

There has been some consolidation in the segmentation space over the years. Equifax acquired IXI, which had earlier acquired Looking Glass and their cohorts segmentation system. Pitney Bowes acquired MapInfo, retiring MapInfo's PSYTE segmentation in lieu of licensing Personicx from Acxiom. Nielsen rebranded Claritas as Nielsen Consumer Activation.

TABLE 9.5   Common US geographic levels

Level	Areas	Average Number of Households
Census		
State	51	2,356,143.2
County	3,143	38,232.0
Tract	73,057	1,644.8
Block Group	217,740	551.9
Block	11,078,297	10.8
Postal		
ZIP Code	30,383	3,965.7
ZIP+4	29,916,069	4.0
ZIP+6	90,000,000 (est.)*	1.3

* A distinct ZIP+6 reflects an actual drop point for mail delivery. Estimating a precise count of residential ZIP+6s becomes more difficult when considering, among other things, how mail is delivered to large apartment buildings, condos, etc. In the majority of cases, however, a single ZIP+6 represents an individual household.

TABLE 9.6   Examples of syndicated segmentations

Company	Segmentation	Granularity (Lowest Level)
Acxiom	Personicx	Household
Equifax (IXI)	Investyles, financial/ economic cohorts	ZIP+4
ESRI	Tapestry	Block Group/ZIP+4
Experian	Mosaic	Household
Neustar	ElementOne	Household/Person
Nielsen Consumer Activation (Claritas)	PRIZM, P$YCLE, ConneXions	Household

The underlying approaches to geodemographic segmentation systems have generally not changed much over the years, optimizing on dimensions such as affluence, family composition and age. As one might expect, additional emphasis has been placed on optimization techniques and machine learning. Although the US Census and related products, such as the American Community Survey,[2] still provide a foundation for timely and accurate universe updates, the census no longer drives segmentation models as it once did. Third-party providers of demographic data continue to gain dominance in supporting household- and person-oriented solutions.[3]

Syndicated segmentation systems have traditionally been robust, optimizing characteristics against a wide array of behaviours to provide a general-purpose schema that fits the needs for multiple use cases. More recently there has been an increasing number of industry-specific schemas. For example: Nielsen's P\$YCLE and ConneXions systems are optimized on financial/investment behaviours and technology behaviours, respectively. There has been some demand to expand into other verticals as well (eg automotive, retail, restaurants, media).

There has also been desire in the industry to move away from self-reported behaviours collected via traditional or online surveys. Respondent recall bias, among other issues, can lead to inaccurate measures of behaviours. Alternatively, big data sources offer access to actual behaviours.[4] Respondent recall is not an issue as big data is often passively collected.

This is not to say that these data sources are challenge free. Unlike demographics, there is no source of behavioural data that reflects the total US population. Big data contains extensive information – often times at a more granular level than a typical survey can measure – but can suffer from coverage and data completeness issues. Cable companies distribute television content to US households and can license tuning data to market researchers. However, the data content will be restricted to the cable company's subscriber base and only to those devices that are return-path capable. Similarly, consumer packaged-goods retailers commonly use loyalty card programmes to encourage loyal buying behaviours, but this data will not reflect purchases made by non-cardholders or trips where the cardholder did not have their loyalty card scanned.

Data availability can also pose issues. Contractual or legal arrangements may result in limited use of the data. It may come pre-aggregated or with certain sectors of the data omitted, which prevents a comprehensive analysis of the dimension of interest. Licensing costs can also prove a roadblock to procuring and integrating this data into segmentation systems.

Perhaps the most striking aspect of change has been in privacy. Concerns over privacy have resulted in the loss of some traditional data sources (eg drivers' licence and vehicle registration data). Leveraging non-traditional sources in finer segmentation models is complicated, however, as a grey area exists between what people are and are not willing to reveal about themselves. People may divulge personal information in an online space provided it is by choice (eg Facebook, blogs, website account registrations) but will opt out of tracking services where such behaviours are monitored in the background (such as with Do Not Track, which aims to prevent web applications from tracking individual users). From an advertising perspective, this means that consumers are exposed to advertisements marketed at the lowest common denominator instead of receiving an ad tailored to their own specific preferences and interests. This disadvantages all parties involved, which raises the question: is there a sweet spot where consumers are willing to reveal *some* information about themselves in order to receive better choices? For example: capturing behaviours while guaranteeing that personally identifiable information is masked may alleviate privacy concerns yet still enable more informed marketing decisions.

Current solutions to these questions vary and can be partially attributed to more relaxed privacy guidelines in the United States, as compared to Europe. One approach is to use only household or individual data acquired with direct consumer permission. Another is to use aggregate-level data, but at the time of writing there is currently no consensus specifying the minimum level of granularity required to alleviate privacy concerns. This issue exposes the opposing forces of contemporary geodemographic data usage: the desire to dive deeper into geographies – and to reach specific households or individuals – is limited to the extent that gathering information and developing models at the same level are viewed as intrusive. This struggle will continue to be a core issue for the industry going forward.

Conclusion

In this chapter, we have seen that the two main prerequisites for geodemographic segmentation in a country are the availability of census-type data for small areas, and the ability to locate the areas geographically 'on the ground'.

Therefore, segmentation systems are more likely to have been created for developed countries such as in Europe, North America and Asia/Pacific, rather than developing countries such as in Africa and parts of South America.

The market for geodemographics in the United States is served by segmentations at a variety of levels, from small areas down to households and persons. However, privacy concerns have resulted in the loss of some traditional data sources, and the industry has yet to work out a balance between 'intrusive' use of individual data, producing well-tailored advertising, and less specific targeting producing more generic advertising.

Notes

1 Counts for postal codes represent residential codes only – businesses and other types are omitted.

2 The American Community Survey is a large annual survey of over two million households per year. It was designed to replace some of the detailed social questions asked in the census. Annual data is available for high-level geographies (areas over 65,000 people) and on a running five-year basis for low-level geographies (eg block group).

3 There are many third-party demographic data providers in the United States. Examples include Acxiom, CoreLogic, Epsilon, Equifax, Experian and TransUnion.

4 Big data and other forms of non-traditional data frequently come from third parties. Data is typically captured or generated as a by-product of their primary business and is then repurposed for other uses.

CREATING YOUR OWN INFORMATION PRODUCTS

10

Introduction

The geodemographics market in the UK is well served with products – in the region of 40 off-the-shelf classifications and discriminators are available, for a variety of purposes and levels of geography. These ready-made systems have been well designed to meet clients' needs and invariably users will look no further. And yet, in some situations, a bespoke solution may be called for – creating your own information product, rather than using the same solution as others.

The main aim of this chapter is to discuss bespoke approaches to building your own geographical information products, covering:

- Why would a customized solution be appropriate?
- The options for building a bespoke geodemographic discriminator.
- Approaches for estimating market sizes at neighbourhood level.

Why build a customized solution?

A bespoke development, by definition, allows the product to be tailored to meet the needs of your business and/or your industry sector. Some situations calling for a customized solution would be:

- If you require segmentation based on different sources, for example using your company's own customer data or inputs specific to your industry.
- If you want the segments to be fully aligned to your business needs, in terms of their definitions and/or their descriptive labels.
- If you need to have estimates of consumer demand at neighbourhood level that are more accurate than those available off the shelf.

The first two situations may either imply building a bespoke discriminator or customizing one of the existing products, respectively. The third situation may imply developing a bespoke set of small-area estimates for your market.

> **KEY POINT**
>
> One possible reason for a customized solution that is sometimes given by users is: 'I don't want to be contacting the same consumers as my competitors.' Generally, this justification is insufficient – two companies in the same sector might be using the same system; however, in all likelihood they will be applying it in different ways and not necessarily targeting the same segments. So it is usually more about what you do with a segmentation that matters, rather than having your own unique segmentation.

Building a bespoke discriminator

In the early days of geodemographics, building a neighbourhood classification was a sizeable project, taking weeks or months, and involving a significant amount of mainframe processing. Nowadays, of course, the same analysis can be carried out in a fraction of the time, on a moderately powerful computer.

Further, the open data movement (discussed in Chapter 3) has been extended to 'open geodemographics'. This means that the 2011 Output Area Classification (OAC) (described in Chapter 4) has been released together with the programs and data that were used to build it. The classification was developed using R, a widely used open-source language for statistical computing, and all of the analytical steps are available as files of R code.

These advances allow analysts to create their own bespoke versions of OAC, such as the London Output Area Classification (LOAC) and the Temporal Output Area Classification (TOAC), which looks at geodemographic change between 2001 and 2011 (for further information, see **http://www.opengeodemographics.com/**).

Even though the data and software may be freely available to build a bespoke discriminator, the analyst's thought processes cannot be automated or ignored. At each step in the process, results need to be interpreted and decisions taken – these are the skilled and time-consuming tasks in geodemographic development. For this reason, a user should have a strong reason or a sound business case before deciding to build their own discriminator.

Hands-on practical

A hands-on practical has been created to accompany this book, so that readers may experience the process of building a geodemographic classification for themselves. The practical is accessible online and may be run as supplied or used as a basis for further analysis. It uses a set of materials that have been designed and created by Dr Luke Burns from the University of Leeds. Dr Burns specializes in quantitative geography and regularly delivers analytical computing courses in this field. In the section below he introduces the aims of the hands-on practical and the approach taken.

Hands-on practical task: building your own geodemographic system

Dr Luke Burns, *University of Leeds*

Now that you have read about the theory and application of geodemographics and, more importantly, some of the decisions that must be taken along the way when building any such system, the online practical exercise described in this section will give you a hands-on opportunity to apply this knowledge using real data. This section builds upon the material covered thus far in this book, in particular the development process described in Chapter 4, and offers a framework through which you can create your own simple general-purpose geodemographic system. This is not designed to be exhaustive nor will it make you an expert but it will show you the steps that you will be required to navigate should you need to build a system in the future.

To make this exercise as accessible as possible, principally open-source data and software are used. Alternative packages will also be discussed should you have other software or wish to take a different approach, such as SPSS or MapInfo. Specifically, you will be introduced to the following freely available tools and data:

- the UK Data Service (a free online repository of UK census and boundary data);

- open-source statistical software (such as PSPP);

- open-source GIS (such as QGIS).

The exercise focuses on the construction of a small geodemographic system for Leeds, UK. Leeds is a northern UK city with a population of just over 750,000 (2011 census). It has a particularly diverse population with regards to certain variables whilst also sitting very close to the national average for others, therefore making it an ideal case study for this demonstration. You will work with a broad selection of UK census data ranging from age and ethnicity to property type and car ownership. The final result will be a simple classification system that profiles small geographical areas based on their characteristics.

During this exercise, you will be guided through the following phases of system development. Each phase will be discussed and hands-on opportunities presented for you to work through. These phases link closely to those set out in Chapter 4:

1 business understanding;

2 selecting the data (input variables and geography);

3 pre-processing the data;

4 clustering the data;

5 labelling, interpreting and visualizing the output;

6 evaluation/application.

You will learn how to source and download UK Census data (for 2011 and earlier), locate and download area boundaries (for mapping), prepare data for classification, run a k-means classification algorithm and visualize the output on a map of Leeds. There will also be an introduction to analysing the composition of each cluster enabling a name or 'pen portrait' to be assigned, which is particularly useful for marketing purposes when trying to target a particular type of consumer.

Many of the considerations discussed in Chapter 4 are evidenced in this exercise, including testing for multicollinearity (correlation), determining polarity and ensuring consistent (standardized) data. Detecting skewness in data and considering variable weights will also be discussed but not necessarily implemented in this hands-on task. You will also gain practical experience of dealing with the modifiable areal unit problem (MAUP) and ecological fallacy as introduced in Chapter 6.

Although this framework is discussed in relation to a case study area of Leeds, UK, the phases you will explore in this exercise can be readily applied to other locations with different data. This therefore makes the practical a nice 'go to' resource for times when assistance is needed to build a geodemographic system.

Full details and resources for the exercise can be found on the accompanying website – please visit either: **www.koganpage.com/product/geodemographics-for-marketers-9780749473822** or **www.barryanalytics.com/geodems4marketers**.

Note that this exercise is very step-by-step but it does assume some prior knowledge of general pc skills and navigation.

Other options for building a discriminator

The increasing availability of big data, particularly to companies such as retailers or mobile phone operators with large numbers of customers to manage, could be perceived as one driver for creating bespoke discriminators. For example, a mobile phone company could summarize the data it holds on customer calling behaviour, and could derive unique behavioural variables at small-area level, which it could use to build its own geodemographic classification. However, the caveat here is that those variables would only measure the activity of that company's customers and would omit the rest of the mobile market. Therefore, the resultant classification would not be truly representative of the market – it might give useful insights about localities with strong customer presence, but would be of less use for areas dominated by competing operators.

As Kyle Poppie and Dave Miller explain in Chapter 9, the US geodemographics industry faces exactly the same issues around how to use big data, when behaviour data really needs to be market-wide – the data availability issues that they discuss apply equally to the UK.

In view of the time and effort required in building a geodemographic discriminator, several of the UK suppliers have created a shortcut option. Each has developed its own low-level classification, containing several hundred 'building-block' types, which may be flexibly grouped together to form a customized classification. External inputs, such as market research data, may be employed to cluster the building blocks in order to discriminate well for the specific market or sector. The outcome is a new purpose-built segmentation – the segments are finally interpreted and labelled accordingly.

This shortcut method has much to commend it – shortened development time, inclusion of external data and the opportunity to interpret and label the resultant segments in a relevant way for that market. The only obvious disadvantage is the use of an off-the-shelf set of building-block types, rather than starting the development from the most discriminatory input variables for the market in question. The question of how much difference this would make is one that remains to be answered.

Small-area estimation

Most suppliers of geodemographic systems will also provide a range of other datasets, including small-area estimates of consumer demand for various domestic markets such as furniture, packaged groceries, wines and spirits, and so on. For each market, the demand estimates are typically produced by applying a research profile (often obtained from one of the sources in Table 6.4) to the population profile of each area, using their neighbourhood classifier to link the two sources (see Chapter 6 for further discussion).

An alternative approach is to produce bespoke small-area estimates, driven by the appropriate source of research on your market. For many markets, purchasing rates vary demographically – for example, they may depend on factors such as age, social class, working status and so on. Provided that this is the case, then it becomes possible to obtain more accurate demand estimates using census data than from geodemographic profiles.

The following case study demonstrates how this was achieved in the context of the 'eating out' market.

CASE STUDY   Development of small-area demand estimates for the 'eating out' market

A restaurant retailer wanted to obtain a set of demand estimates for 'eating out' meals in order to provide an accurate input to its site location analysis system. The retailer operated a number of brands in different sectors of the market and required estimates for seven sectors, ranging from fish and chips to formal dining. The estimates were developed in a series of stages, which are summarized in Figure 10.1, and are outlined below.

Stage 1: market research survey
The company had access to a market research survey for this market that measured the number of meals eaten in the last week, for a sample of adult individuals, together with their demographics.

Stage 2: model demand by demographics
A regression model was developed for each sector, to estimate the rate of eating out based on demographics that were also available in census outputs – characteristics such as age group, presence of children, car ownership, working status, social grade and region. The models were significantly more discriminatory than conventional geodemographic methods, thus justifying continuation to the next stage.

Stage 3: apply model to census microdata
The scorecards were applied to the file of census microdata for individuals in the UK Census, using the characteristics included in each model. This

FIGURE 10.1 Development of demand estimates for 'eating out' meals

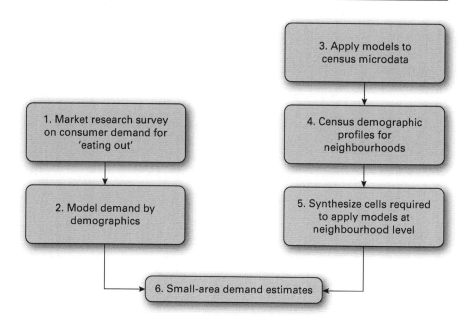

generated an estimated demand, by sector, for each adult in the dataset. These results were summarized at local authority level, giving a first set of estimates that could be mapped and sense-checked.

Stage 4: census demographic profiles for neighbourhoods
The small-area census statistics were extracted for the demographics required by the models; in some cases, the required demographic variable had to be estimated, with the help of census microdata. For example, the number of adults with children was required for some of the models; however, census output gave only households with children.

Stage 5: synthesize cells required to apply models
In order to apply the models to census neighbourhoods, cell counts were required for all combinations of demographics used for each model. The individual census microdata was employed again to synthesize a set of cell counts, which reflected the relationships between the variables and were consistent with the demographic profiles obtained in stage 4.

Stage 6: small-area demand estimates
The demographic models could then be applied to the synthesized cell counts
for each neighbourhood, to obtain a set of market estimates. Finally, the
market estimates were scaled up to national levels, for consistency with the
known total market size for each sector.

The outcome from this development was a set of accurate demand estimates that
could be employed in the retailer's site location system. There were other spin-off
benefits, for example the project confirmed the importance of social grade as a
discriminator across these sectors and delivered a set of social grade profiles for
all census neighbourhoods.

Conclusion

This chapter has shown that:

- Building your own geodemographic discriminator should not be
  undertaken lightly, but it allows you to include different sources to
  create a market-specific product and describe the segments to
  meet your business needs.

- By starting from the building-block segments held by some
  geodemographic suppliers, the development time and budget may
  be reduced.

- For markets where demand is dependent on standard
  demographics and a good market research source is available,
  more accurate small-area demand estimates may be created via
  analysis and modelling.

Finally, the hands-on practical introduced in this chapter provides an
opportunity for interested readers to experience the process of building a
geodemographic classification for themselves.

LOOKING TO THE FUTURE

11

Introduction

Geodemographics in the UK has been thriving for the best part of four decades, but what are its prospects for the future? Will big data take over from neighbourhood analysis? Or is there still a role for neighbourhood contextual effects, as Richard Webber argued in Chapter 2? This chapter looks to the future, identifies likely trends and offers the author's personal predictions.

The geodemographics industry has done well over the decades to harness advances in technology and new information sources as soon as they have come 'on stream'. Its prospects for the future will primarily be linked to the availability of population characteristics at small-area level – if these were heavily curtailed, there could be serious implications for geodemographic segmentation. Therefore, our future gazing focuses on key sources – census, open data, administrative files and big data.

The main thrust of this chapter will be to discuss developments in each of these sources, which could potentially impact on the geodemographics industry. We start with the census – the main focus will be on England and Wales and developments from the ONS, as these are likely to indicate the general direction of censuses in the UK.

Before discussing data supply, however, let's consider how the demand for geodemographic information is likely to change in the future.

Likelihood of change in user demand

In the early years of the industry, the direct marketing sector made heavy use of geodemographic discriminators for targeting their direct mail campaigns. At that time, users constantly called for the area units to become smaller and more accurately defined, which led to postcode-level segmentations and ultimately to individual-level products.

Despite this trend, user demand for neighbourhood classifications continues to remain strong and is unlikely to weaken in the future. There are several strong factors underpinning postcode and OA systems. First, from a data protection perspective, area classifiers are less problematic than their individual-level counterparts. Individual-level systems generally have to access personal data in order to apply it, whereas area products are driven entirely by postcodes, which are both simpler and safer. These considerations apply to both customer databases and to market research surveys. Indeed, the MRS Code of Conduct requires that identities of respondents are kept confidential, which would limit the general use of individual-level geodemographics.

Second, many users require information in order to plan the provision of stores and services to local area populations, and individual-level discriminators are unnecessarily granular for this purpose. The demand for geodemographic classifiers at small-area level particularly comes from the retail sector, for the kinds of applications discussed in Chapter 7.

Given the continuing demand for neighbourhood classifications, the following sections discuss developments in data supply, focusing on four main source types – census, open data, administrative data and big data.

Census developments

Recent history

From the early 2000s, the ONS started to explore new ways to count the population in order to overcome the drawbacks of the decennial 'big bang' census. Large-scale administrative databases were well established by that time and were being increasingly used for statistical purposes. First thoughts on how to change the system came in 2003, with the ONS plan for an Integrated Population Statistics System (see ONS, 2003).

Another milestone was the Treasury Select Committee report in 2008 on 'Counting the Population', which concluded that the traditional census had almost had its day. Amongst its findings, the report recommended that the UK Statistics Authority set objectives to ensure that data currently gathered throughout the UK (eg administrative data) could be used to produce annual population statistics, to enable 2011 to be the last-ever traditional census.

Soon after the general election in 2010, the new government added to the pressure for change, saying that: 'The Census, which takes place every ten years, is an expensive and inaccurate way of measuring the number of people in Britain. Instead the government is examining different and cheaper ways to count the population more regularly using existing public and private databases' (Frances Maude, Cabinet Office minister, quoted in the *Daily Telegraph*, 9 July 2010).

The ONS Beyond 2011 Programme had been set up in 2009 to research alternative approaches to producing population statistics and recommend the best way forward. This major research project was fully launched as soon as the 2011 census fieldwork was completed, and ran for three further years.

Beyond 2011 programme

In the early stages of the programme, ONS looked at the different ways in which the production of population statistics and small-area demographic profiles was approached in other countries. Eight initial options were identified for further research – these are shown in Figure 11.1 and were explained by ONS (2013). The options broadly fell into three groups:

- Census options, which are variations of the current decennial census approach.
- Administrative data options, which estimate population size using data from administrative sources, and conduct annual surveys to collect data on population characteristics.
- Survey option, which uses a high-quality address register and carries out annual surveys to estimate population size and characteristics.

By October 2013, ONS had narrowed down the options to two potential approaches: 1) a census once a decade, like that conducted in 2011, but carried out primarily online; 2) a census based on administrative data combined with an annual survey of 4 per cent of the population.

FIGURE 11.1   Beyond 2011 – full range of approaches assessed

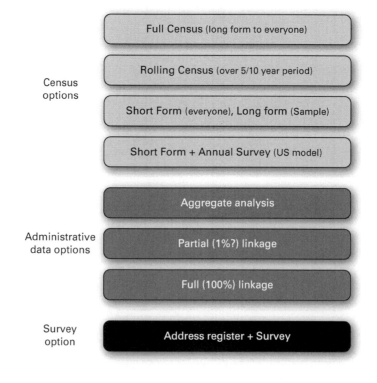

ONS researched the second approach in greater detail, using data from administrative sources such as the NHS patient register and the Department for Work and Pensions/HM Revenue and Customs customer information system to produce indicative population estimates and check their accuracy.

In autumn 2013, ONS carried out a full public consultation, asking users for their views on these two approaches. ONS clearly demonstrated the potential of the second approach, as a technique for producing population estimates from existing administrative sources. At the same time, ONS explained that survey results would need to be accumulated over a number of years, in order to produce small-area data on population characteristics. This aspect of the second approach was of particular concern to many users, including those requiring small-area statistics for geodemographic analysis.

The response from users contained some consistent messages:

- There was continuing demand from all sectors for detailed information about small areas and small populations as offered by the decennial census.

- There was a strong concern that the proposed use of an annual sample survey (in the second approach) would not meet these needs.

- The more frequent statistics that could be provided between censuses by the use of administrative data and annual surveys would be welcomed, but not at the expense of the detailed statistics.

- Whilst the methods using administrative data and surveys showed considerable potential, there was concern that these were not yet mature enough to replace the decennial census.

- Many respondents proposed a hybrid approach, making the best of both approaches, with an online census in 2021 enhanced by administrative data and household surveys.

ONS also commissioned an independent review of methodology led by Professor Chris Skinner. The findings from this review (see Skinner, Hollis and Murphy, 2013) attached low risk to the online census approach and higher risk to the administrative data approach. The review recommended further research into the use of administrative data and that this approach was not yet ready to replace the decennial census.

In March 2014, the National Statistician published her recommendation to the board of the UK Statistics Authority (UKSA) for:

- an online census in 2021, with support for those who are unable to complete the census online;

- increased use of administrative data and surveys, in order to enhance the statistics from the 2021 census and improve annual statistics between censuses.

The National Statistician recommended that this would make the best use of all available data to provide the population statistics required, and would offer a springboard to greater use of administrative data and annual surveys in the future.

The UKSA board accepted and endorsed the National Statistician's recommendation and commended it to the Cabinet Office. The government response, in July 2014, agreed with the recommendation for an online census in 2021, together with increased use of administrative data. At the same time, the government expressed the ambition that censuses after 2021 would be conducted using other sources of data and would provide more timely statistical information.

Therefore, the UK government's aim is that the final full census will take place in 2021. To enable this to happen, ONS is encouraged to conduct parallel research during the period up to 2021, using administrative data and sample surveys. This dual running should be designed to validate the feasibility of that approach, so that the alternatives can be evaluated again following the 2021 census.

Census Transformation Programme

The ONS Census Transformation Programme began in January 2015, in order to deliver to the above objectives. There are three main strands to the programme:

1 *2021 online census operation*
The main aim of this strand will be to research, develop, implement and conduct the 2021 UK Census. Although online response was available in 2011 and 16 per cent of households took this option, the majority still used paper forms. The 2021 census will aim for the

primary channel to be online, but will still need to collect data from the 'digitally excluded' and those who prefer other modes of completion. Therefore this strand will need to implement a multimode census, taking into account the implications of primarily online data collection.

At the same time, this strand will look at modernizing field processes, including the technology used by the census field force and the options for e-mail contact and postal reminders.

In addition, ONS will be aiming to make greater use of administrative data in the data collection process, including an enhanced address register for operating the census and systems for field force management and targeting non-response follow-ups. At the time of writing this book, ONS is planning to use the Ordnance Survey's AddressBase product (see below) for the address register, supplemented by additional addresses for communal establishments.

2 *Integrated outputs*

The second strand will focus on integrating census, administrative and survey data to produce more detailed and more frequent outputs. A range of options will be researched, from ad hoc data integration through to full continuous integration of all sources.

This strand will aim to produce enhanced census and/or population statistics; it will provide research results in order to deliver early benefits and obtain feedback from users.

3 *Beyond 2021*

The third strand will develop new methods using administrative data and surveys, which will ultimately be evaluated against 2021 census outputs. Research statistics will be produced each year and will be reviewed with users.

By 2021, the aim will be to have covered as much of the breadth, detail and accuracy of census outputs as possible. After validating the results against the 2021 census, proposals will be produced for the future of the census and population statistics.

The future of the UK Census – summary

The UK Census will continue to deliver small-area population estimates and characteristics, required by the geodemographics industry, at least through to the 2020s. The last full population count may take place in 2021, by which time a viable alternative approach for 2031 will need to have been developed and proven. However, this remains to be seen, as it seems like an ambitious goal – particularly in the absence of a UK population register.

Open data developments

The open data movement is already well advanced in the UK, and seems likely to continue strongly – thanks largely to organizations such as the Open Data User Group (ODUG) and the Open Data Institute (ODI).

At the time of writing, a major step forward seems likely to be taken with the creation of a National Information Infrastructure (NII) containing all public-sector datasets that will be 'open by default' for all to use. Early in 2015, ODUG published a paper explaining the importance of the NII and calling on the government to plan this development so that it delivers real value (see Open Data User Group, 2015).

Open addresses

Possibly the most important data source that has yet to be made open, in the UK, is the list of all addresses in the country. An open national address list would be beneficial for all companies that deliver goods and services to consumers.

At present, there are two principal sources of addresses nationally – the Postcode Address File (PAF) from Royal Mail and the definitive National Address Gazetteer (NAG) from GeoPlace. Ordnance Survey offers a series of AddressBase products based on these sources and, as mentioned above, AddressBase will form the core of the register to be used by ONS for the 2021 census.

At the time of writing, none of these products is open – the private sector may only access address information by paying licence fees, which are uneconomic for smaller companies to bear.

At the start of 2014, the Department for Business Innovation and Skills published a review, written by Katalysis Limited, on the need for an open National Gazetteer (BIS, 2014a). The review recommended that a basic address list should be freely available for all users, while premium products should continue to be sold. The responses to this review were published in April 2014, and showed that users strongly supported this recommendation (BIS, 2014b).

Later that year, the ODI took the initiative to launch a project that will create an open address list. The ODI set up a company, OpenAddresses, dedicated to this purpose. OpenAddresses is building a new list, by a combination of methods – primarily inferring addresses through computerized analysis of existing open sources, and adding in addresses obtained through crowdsourcing. While this initiative is to be congratulated, it can never achieve the 100 per cent coverage required by the geodemographics industry. Therefore, one must hope that the project will demonstrate to the government that the definitive National Address Gazetteer needs to be made open and that steps will be taken for this to happen.

Administrative data developments

Earlier in this chapter we considered the potential for deriving additional value from administrative data, in the context of the work done by ONS to produce trial population estimates. This application is just one of many possible secondary uses of administrative data – others include deriving annual measures of health and deprivation using data on the numbers receiving incapacity benefits and unemployment statistics, rather than relying solely on the decennial census statistics.

The potential for increased research based on administrative data was identified by the Administrative Data Taskforce report at the end of 2012 (UK ADRN, 2012). This led to the formation of the Administrative Data Research Network (ADRN), which was launched in 2014. The main aim of the ADRN is to increase the use of linked de-identified administrative data for research, in order to achieve value for money and additional benefits from data that is already being captured for other purposes.

The ADRN is designed to enable accredited researchers to submit proposals for research projects using administrative data. Initially, the service is only available to those in the academic sector or working for a government department. At the time of writing, an extension of availability to the third sector (the voluntary sector) is under consideration. We hope that this will be further reviewed and extended to all other sectors, as the ADRN ought to be accessible to anyone wishing to use the data to derive added value for the public good.

Research proposals are submitted to the ADRN and reviewed by an independent approvals panel. Once a proposal has been approved, the ADRN negotiates with the owner of the requested data in order to obtain the necessary dataset(s) on behalf of the researcher.

There are four administrative data research centres across the UK, which take responsibility for processing the data and providing access to researchers. Confidentiality is paramount – this is achieved by ensuring that, other than the data owner, no one has access to both personal identifiers and attributes about people. The researcher receives access to a linked de-identified dataset, in a secure laboratory environment.

At the time of writing, the ADRN is still in its early days and benefits have yet to be seen. There seems to be no doubt that the initiative will lead to increasing use of administrative data. We can only hope that access will be extended to include the private sector, and that all sectors will be able to research, analyse and apply administrative data for the public good. Just as academic researchers are starting to create and test new datasets based on administrative databases, commercial analysts could, for example, explore the potential of the data for constructing wealth indices and new measures for geodemographic analysis.

Big data developments

Organizations such as banks, retailers, phone companies and utilities need to store vast amounts of data in order to operate their customers' accounts. These datasets are popularly known as 'big data' and were discussed in Chapter 2.

Commercial data analysts have been mining big data for many years, employing it to segment and target customers. The most significant trend from the early 2000s onwards has been the explosion in data volumes, due primarily to the growth in internet data. This has led to new technologies for distributed processing and advances in computer languages, requiring different skill sets and resulting in new job roles – such as the 'data scientist'. This trend will continue, at an ever-increasing rate, fuelled by further developments such as the take-off of radio-frequency identification (RFID) chips and the internet of things.

Academic researchers are becoming involved in big data analytics in a more systematic way. In the UK, this is taking place through the ESRC's

Big Data Network initiative, which was set up as three phases in order to help optimize the use of big data. The ADRN, outlined above, is the first phase of this initiative, while the second phase is to focus on business and local government data via centres such as the Consumer Data Research Centre (CDRC). The CDRC is run as a partnership by researchers from University College London, together with the universities of Liverpool, Leeds and Oxford. Analysts from these centres carry out research projects in collaboration with commercial sponsors. The third phase will focus on social media and third-sector data – at the time of writing, this phase is still at an early stage of development.

It is natural to think that big data could provide additional useful sources for geodemographic systems in the future. The main drawback was pointed out by Poppie and Miller in Chapter 9 – while each big dataset gives a huge amount of detail, it is limited to those customers who have purchased products or hold accounts with that organization and so it cannot yield data on the entire population, or even a representative sample from it.

One exception occurs in the utilities sector – virtually every household consumes electricity and so there could be an opportunity to collect together data from energy suppliers and use it to derive new social indicators. Furthermore, by the early 2020s, customers are due to be switched over to smart meters that capture energy consumption throughout the day, at 30-second intervals, and so will be able to give a detailed usage profile for each household.

The 'Census 2022' project at Southampton University began in 2013 to conduct research into the use of these consumption profiles to predict characteristics such as household size (larger households consume more energy) and presence of children (households with children have different energy patterns across the day). The project's aim is to explore the usefulness of smart-meter data for predicting various household attributes, using samples of metered households for which demographics and other characteristics are also known. At the time of writing, this work is still at an early stage, but should help to show the potential of big data on smart-meter usage, hopefully leading on to larger-scale development projects. This source could provide one component of a system of population estimates based on administrative sources and big data that may help obviate the need for a census in 2031.

The ONS Big Data Project was launched in 2014 with goals to develop an understanding of big data within ONS, demonstrate its potential within official statistics and investigate the methodological and technical issues. Early projects have included analysing Zoopla data on property rental prices and using geo-located Twitter records to gain new insights into student mobility.

Conclusion

In this chapter, we have seen that:

- The demand for geodemographic information remains strong and is likely to continue for the foreseeable future.
- With the decision to hold the next census in 2021, the geodemographics industry in the UK has a secure primary data source at least through to the end of the 2020s.
- Users may expect to see and examine a new series of research estimates based on administrative sources during the run-up to the 2021 census. These developments should help to improve the accuracy of small-area population estimates and so are to be welcomed.
- Open data developments are likely to continue; we are still some way off from having an open national address database, which would be greatly welcomed by all users.
- The academic sector is becoming involved in big data analytics, in a systematic way. Over time, this should help to optimize the use of sources such as administrative files, business databases and social media data.
- Further research and development are needed before customer databases can contribute to geodemographics, although early work using smart-meter data has demonstrated the potential to correlate household demographics with energy consumption.

APPENDIX A
Useful information sources

UK census offices

Each of the three census offices provides guidance and methodology on its website, including contact details for its Customer Services department and links for downloading census tables:

For England and Wales, visit the Office for National Statistics (ONS) website: **http://www.ons.gov.uk/ons/guide-method/census/index.html**.

The ONS is also responsible for disseminating census output for the UK as a whole, gathered from all three census offices – see: **http://www.ons.gov.uk/ons/guide-method/census/2011/uk-census/index.html**.

For information about the census in Scotland, visit the National Records of Scotland (NRS) website: **http://www.scotlandscensus.gov.uk/**.

For census data on Northern Ireland, see the Northern Ireland Statistics and Research Agency (NISRA) website: **http://www.nisra.gov.uk/Census.html**.

Neighbourhood statistics

The ONS also operates the Neighbourhood Statistics website, which enables users to obtain statistics about an area or a neighbourhood summary report for any location in England and Wales – see: **http://www.neighbourhood.statistics.gov.uk/dissemination/LeadHome.do**.

Open data

The Open Data Institute (ODI) was set up to promote and extract value from open data. It runs regular events and training courses, and its website includes case studies and membership details – see: **http://opendatainstitute.org/**.

The main aim of the Open Data User Group (ODUG) is to encourage the release of open data through the government's Public Sector Transparency Board – see: **https://www.gov.uk/government/groups/open-data-user-group** [accessed 22 October 2015]. NB, in 2015 the Open Data User Group ceased to operate, following its initial three-year term. It remains to be seen whether ODUG will be continued.

Catalogues of existing open datasets are available at: **http://data.gov.uk/**. This website includes a request process enabling users to ask for specific datasets to be released as open.

The Open Geodemographics website is a portal to existing open neighbourhood classifications in the UK. See: **http://www.opengeodemographics.com/**.

The following website provides an interactive map of the UK for displaying open geodemographic classifications: **http://public.cdrc.ac.uk/**.

The website of the Output Area Classification User Group is located at: **https://plus.google.com/communities/111157299976084744069**.

UK Data Service

The UK Data Service is a comprehensive resource that holds social and economic data on behalf of the UK census offices, government departments, intergovernmental agencies, research institutes and researchers. Datasets may be accessed by researchers and students in the academic sector, government analysts, business analysts and others. For further information see: **http://ukdataservice.ac.uk/**.

Administrative Data Research Network (ADRN)

The ADRN provides a bespoke service that enables researchers to carry out economic and social research using administrative data. See: **http://adrn.ac.uk/**.

Consumer Data Research Centre (CDRC)

The CDRC is a multi-institution laboratory for analysing consumer-related data from around the UK. See: **http://cdrc.ac.uk/**.

UCL Centre for Advanced Spatial Analysis (CASA)

CASA is a department of University College London that focuses on applying geospatial analysis, modelling and visualization techniques for use in city planning. For examples of its work, see: **http://www.bartlett.ucl.ac.uk/casa**.

DataShine

DataShine is a mapping platform developed by researchers at UCL. A series of DataShine websites has been created for visualizing large datasets, including the 2011 census and OAC. See: **http://blog.datashine.org.uk/** [accessed 18 November 2015].

Geodemographics Knowledge Base (GKB)

The Geodemographics Knowledge Base (GKB) is a comprehensive directory of websites on census and other data sources, geodemographics, geospatial analysis and open data. Its Europe section provides international sources of official statistics and other useful sites. To visit the GKB, go to: http://www.geodemographics.org.uk/.

Social grade

The Market Research Society publishes a very useful guide, known as *Occupation Groupings*, which is designed to help research practitioners determine the social grades of respondents (see MRS, 2010).

The guide may be obtained direct from MRS – for further details, see: **https://www.mrs.org.uk/intelligence/occupational_groupings**   [accessed 14 October 2015].

APPENDIX B
The structure of a UK postcode

All postcodes in the UK follow a standard structure, which is explained in this appendix using the example below.

Take the postcode:

MK	17	8	TA
Area	District	Sector	Unit

The postcode system is hierarchical, the top level being the postcode area comprising one or two alphabetic letters, eg 'MK' represents the Milton Keynes postcode area.

The second level is postcode district, eg 'MK17'. This level is also commonly known as the 'outcode' and usually provides sorting-office routing information for Royal Mail. MK17 covers a number of villages south of the town of Milton Keynes.

The third level is postcode sector, eg 'MK17 8' – postcode sectors are sometimes used as a convenient geography for marketing purposes, including catchment area definition and planning marketing activities.

The fourth and final level is the full unit postcode, eg 'MK17 8TA' for Church Road, Woburn Sands. The combination of sector and unit (the '8TA' part) is often called the 'incode' and may be used by the delivery offices to get the mail on the right vehicle for final delivery.

Whilst the 'incode' part of the postcode is always one numeric character followed by two alphabetic letters, the 'outcode' can have several different formats and be anywhere from two to four alphanumeric characters long.

The numbers of postal codes in the UK, as at Q3 2014, were as follows:

Geographic Level	Count
Areas	124
Districts	2,987
Sectors	11,185
Postcodes	1,744,693

SOURCE: BPH Data Ltd (http://www.bph-postcodes.co.uk/guidetopc.cgi)

REFERENCES

ABS (2011) [accessed 31 July 2015] 1270.0.55.001 – Australian Statistical Geography Standard (ASGS), Volume 1, Main Structure and Greater Capital City Statistical Areas [Online] http://www.abs.gov.au/ausstats/subscriber.nsf/log?openagent&1270055001_july%202011.pdf&1270.0.55.001&Publication&D3DC26F35A8AF579CA257801000DCD7D&July%202011&23.12.2010&Latest

Allo, N (2014) A challenge for geomarketing in developing countries, *International Journal of Market Research*, **56** (3), pp 297–316

Bermingham, J, Baker, K and McDonald, C (1979) The Utility to Market Research of the Classification of Residential Neighbourhoods, MRS Conference. Reprinted (1997) in the *International Journal of Market Research*, **39** (1)

BIS (2014a) An Open National Address Gazetteer, Katalysis Limited, Department for Business Innovation & Skills

BIS (2014b) An Open National Address Gazetteer: Responses Received, Department for Business Innovation & Skills

Burrough, PA and McDonnell, RA (1998) *Principles of Geographical Information Systems*, 2nd edn, Oxford University Press, Oxford

Butler, T and Hamnett, C (2007) The geography of education, *Urban Studies*, **44** (7), pp 1161–74

Butler, T and Hamnett, C (2011) *Ethnicity, Class and Aspiration: Understanding London's new East End*, Policy Press, Bristol

Butler, T and Robson, G (2003) *London Calling: The middle classes and the re-making of inner London*, Berg 3PL, Oxford, New York

Chapman, P, Clinton, J, Kerber, R, Khabaza, T, Reinartz, T, Shearer, C and Wirth, R (2000) [accessed 31 July 2015] CRISP-DM 1.0 Step-by-step Data Mining Guide [Online] http://www.the-modeling-agency.com/crisp-dm.pdf

Charlton, M, Openshaw, S and Wymer, C (1985) Some new classifications of census enumeration districts in Britain: a poor man's Acorn, *Journal of Economic and Social Measurement*, **13**, pp 69–96

Davies, RL and Rogers, D (eds) (1984) *Store Location and Assessment Research*, Wiley, Chichester

DeMers, MN (2008) *Fundamentals of Geographical Information Systems*, 4th edn, Wiley, Chichester

Dugmore, K, Furness, P, Leventhal, B and Moy, C (2011) Beyond the 2011 census in the United Kingdom: with an international perspective, *International Journal of Market Research*, **53** (5), pp 619–50

Duke-Williams, O and Rees, P (1998) Can census offices publish statistics for more than one small area geography? An analysis of the differencing problem in statistical disclosure, *International Journal of Geographical Information Science*, **12** (6), pp 579–605

Eurostat (2011) [accessed 31 July 2015] Methodologies and Working Papers: EU Legislation on the 2011 Population and Housing Censuses: Explanatory Notes [Online] http://unstats.un.org/unsd/censuskb20/Attachment486.aspx

Flowerdew, R and Feng, Z (1999) The use of fuzzy classification to improve geodemographic targeting, in *Innovations in GIS 6*, ed Bruce Gittings, pp 133–44, Taylor and Francis, London

Fotheringham, AS, Foley, PF and Charlton, M (2008) Automated boundary creation: atomic small areas in Ireland, in *The European Information Society: Lecture notes in geoinformation and cartography*, ed BL Friis-Christensen and H Pundt, pp 241–59, Proceedings of the 11th AGILE Conference on Geographic Information Science, Springer Verlag, Girona

Franks, Bill (2012) *Taming the Big Data Tidal Wave: Finding opportunities in huge data streams with advanced analytics*, Wiley, New Jersey

Furness, P (2004) The international perspective: how censuses vary between countries, in K Dugmore and C Moy (eds) *A Guide to the 2001 Census*, Chapter 14, The Stationery Office, London

Hamnett, C, Ramsden, M and Butler, T (2007) Social background, ethnicity, school composition and educational attainment in East London, *Urban Studies*, **44** (7), pp 1255–80

Harris, R, Sleight, P and Webber, R (2005) *Geodemographics, GIS and Neighbourhood Targeting*, Wiley, Chichester

Heywood, I, Cornelius, S and Carver, S (2011) *An Introduction to Geographical Information Systems*, 4th edn, Prentice Hall, Harlow

Lambert, H and Moy, C (2013) [accessed 31 July 2015] Social Grade Allocation to the 2011 Census [Online] https://www.mrs.org.uk/pdf/Social%20Grade%20Allocation%20for%202011%20Census.pdf

Lazarsfeld, M and Rosenberg, PF (1955) *The Language of Social Research*, Free Press, New York

Lenk, M (2008) [accessed 31 July 2015] Methods of Register-based Census in Austria, *Statistics Austria* [Online] http://unstats.un.org/unsd/statcom/

statcom_09/seminars/innovation/Innovation%20Seminar/
StatisticsAustria_register%20based%20census.pdf

Leventhal, B (1997) An Approach to Fusing Market Research with Database Marketing, *International Journal of Market Research*, **39** (4), October, pp 545–88

Longley, PA, Goodchild, MF, Maguire, DJ and Rhind, DW (2005) *Geographic Information Systems and Science*, 2nd edn, Wiley, Chichester

Martin, D (2002) Geography for the 2001 census in England and Wales, *Population Trends*, **108**, pp 7–15

Martin, D, Cockings, S and Harfoot, A (2013) Development of a geographical framework for census workplace data, *Journal of the Royal Statistical Society: Series A (Statistics in Society)*, **176** (2), pp 585–602

Maslow, AH (1943) A theory of human motivation, *Psychological Review*, **50** (4), pp 370–96

Maslow, AH (1954) *Motivation and Personality*, Harper and Row, New York

McKinsey Global Institute (2011) [accessed 31 July 2015] Big Data: The Next Frontier for Innovation, Competition and Productivity [Online] http://www.mckinsey.com/insights/business_technology/ big_data_the_next_frontier_for_innovation

Meier, E and Moy, C (2004) Social grading and the census, *International Journal of Market Research*, **46** (2), pp 141–70

MRS (2010) *Occupation Groupings: A Job Dictionary*, 7th edn, Market Research Society, London

O'Brien, S and Ford, R (1988) Can We At Last Say Goodbye to Social Class? Market Research Society Conference

ONS (2003) Discussion Paper: Proposals for an Integrated Population Statistics System, Office for National Statistics, October

ONS (2013) [accessed 31 July 2015] Beyond 2011: Options Explained 2, February [Online] http://www.ons.gov.uk/ons/about-ons/who-ons-are/ programmes-and-projects/beyond-2011/what-are-the-options-/index.html

ONS, NRS and NISRA (2012) The Conduct of the 2011 Census in the UK: Statement of Agreement of the National Statistician and the Registrars General for Scotland and Northern Ireland, February 2005, revised July 2012

ONS (2014) [accessed 31 July 2015] Methodology Note for the 2011 Area Classification for Output Areas [Online] http://www.ons.gov.uk/ons/ guide-method/geography/products/area-classifications/ ns-area-classifications/ns-2011-area-classifications/index.html

Open Data User Group (2015) [accessed 31 July 2015] The National Information Infrastructure: Why, What and How? [Online] http://odug.org.uk/

Openshaw, S (1984) *The Modifiable Areal Unit Problem (CATMOG 38)*, Geoabstracts, Norwich

Peucker, TK and Chrisman, N (1975) Cartographic Data Structures, *The American Cartographer*, **2** (1), pp 55–69

Ralphs, M and Tutton, P (2011) [accessed 31 July 2015] Beyond 2011: International Models for Census Taking: Current Processes and Future Developments, *Office for National Statistics* [Online] http://www.ons.gov.uk/ons/about-ons/who-ons-are/programmes-and-projects/beyond-2011/reports-and-publications/early-reports-and-research-papers/index.html

Savage, M and Burrows, R (2007) The coming crisis of empirical sociology, *Sociology*, **41** (5), pp 885–99

Singleton, AD and Spielman, SE (2014) *The Past, Present, and Future of Geodemographic Research in the United States and United Kingdom: The Professional Geographer*, Routledge Informa Ltd, London and New York

Skinner, C, Hollis, J and Murphy, M (2013) [accessed 31 July 2015] Beyond 2011: Independent Review of Methodology [Online] http://www.ons.gov.uk/ons/about-ons/who-ons-are/programmes-and-projects/beyond-2011/beyond-2011–independent-review-of-methodolgy/index.html

Sleight, P (2004) *Targeting Customers: How to Use Geodemographic and Lifestyle Data in Your Business*, World Advertising Research Centre, Henley-on-Thames

Statistics Canada (2012) [accessed 31 July 2015] Boundary Files, Reference Guide, Second edition, 2011 Census, *Statistics Canada* [Online] http://www.statcan.gc.ca/pub/92-160-g/92-160-g2011002-eng.pdf

Statistics Norway (2007) [accessed 31 July 2015] The First Register-Based Census in Norway in 2011: How to Comply with International Recommendations? [Online] http://live.unece.org/fileadmin/DAM/stats/documents/ece/ces/ge.41/2007/sp.6.e.pdf

UK ADRN (2012) Improving Access for Research and Policy: Report from the Administrative Data Taskforce, Economic & Social Research Council

UNECE (2006) [accessed 31 July 2015] Recommendations for the 2010 Censuses of Population and Housing, Conference of European Statisticians, *United Nations* [Online] http://unstats.un.org/unsd/censuskb20/Attachments/CES_2010_Census_Recommendations_English-GUID478c8e0d4a33483381ca030af38fa5b1.pdf

UNECE (2007) [accessed 31 July 2015] Towards the Adoption of EU Legislation for Population and Housing Censuses: Summarising Progress and Highlighting Issues Relevant to Register-Based Censuses, Conference of European Statisticians, Group of Experts on Population and Housing Censuses, *United Nations* [Online] http://live.unece.org/fileadmin/DAM/stats/documents/ece/ces/ge.41/2007/sp.4.e.pdf

UNECE Secretariat (2010) [accessed 31 July 2015] Main Results of the UNECE–UNSD Survey on the 2010–2011 Round of Censuses in the UNECE region, Conference of European Statisticians, Thirteenth Meeting. Geneva, 7–9 July, *United Nations Economic and Social Council* [Online] http://www.unece.org/fileadmin/DAM/stats/documents/ece/ces/ge.41/2010/mtg1/sp.1.e.pdf

United Nations (2008) [accessed 31 July 2015] Principles and Recommendations for Population and Housing Censuses – Revision 2, New York [Online] http://unstats.un.org/unsd/censuskb20/KnowledgebaseArticle10307.aspx

United Nations (2009) [accessed 31 July 2015] Handbook on Geospatial Infrastructure in Support of Census Activities [Online] http://unstats.un.org/unsd/publication/seriesf/Seriesf_103e.pdf

United States Census Bureau (1994) [accessed 31 July 2015] Geographic Areas Reference Manual [Online] https://www.census.gov/geo/reference/garm.html

UNSD (United Nations Statistics Division) (2010) [accessed 31 July 2015] 2010 World Population and Housing Census Programme [Online] http://unstats.un.org/unsd/demographic/sources/census/2010_PHC/default.htm

Valente, P (2010) [accessed 31 July 2015] Census Taking in Europe: How are Populations Counted in 2010, Bulletin *Mensuel d'Information de L'Institut National d'Études Demographiques, Population & Societies*, No 467 [Online] http://www.unece.org/publications/oes/STATS_population.societies.pdf

Webber, R (2009) Response to 'The Coming Crisis of Empirical Sociology': an outline of the research potential of administrative and transactional data, *Sociology*, **43**, pp 169–78

Webber, R (2013) The evolution of direct, data and digital marketing, *Journal of Direct, Data and Digital Marketing Practice*, **14**, pp 291–309

Webber, R and Butler, T (2007) Classifying pupils by where they live: how well does this predict variations in their GCSE results? *Urban Studies*, **44** (7), pp 1229–53

White, I (2014) [accessed 31 July 2015] The 2021 Census within an International Context, *Geodemographics Blog, Geodemographics Knowledge Base* [Online] http://www.geodemographics.org.uk/blog/gkb/ian_white_the_2021_census_within_an_international_context/id/828

Wrigley, N (ed) (1988) *Store Choice, Store Location and Market Analysis*, Routledge, London

INDEX

Note: *Italics* indicate a Figure or Table in the text.